MINERALS, ROCKS, AND FOS

MORE THAN 80 WILEY SELF-TEACHING GUIDES TEACH PRACTICAL SKILLS FROM MATH TO MICROCOMPUTERS, POPULAR SCIENCE TO PERSONAL FINANCE STGS ON BUSINESS AND MANAGEMENT SKILLS INCLUDE:

Accounting Essentials, Margolis
*****Assertive Supervision,** Burley–Allen
Beyond Stress to Effective Management, Gmelch
Business Mathematics, Locke
Business Statistics, 2nd ed., Koosis
Choosing Success: Human Relationships On The Job, Jongeward
Clear Writing, A Business Guide, Gilbert
Communicating By Letter, Gilbert
Communication For Problem-Solving, Curtis
Creative Cost Improvement For Managers, Tagliaferri
GMAT: Graduate Management Admission Test, Volkell
*****How to Read, Understand, and Use Financial Reports,** Ferner
Improving Leadership Effectiveness: The Leader Match Concept, Fiedler
Listening: The Forgotten Skill, Burley–Allen
LSAT: Law School Admission Test, Volkell
Management Accounting, Madden
Managing Behavior on the Job, Brown
Managing the Interview, Olson
Managing Your Own Money, Zimmerman
Meetings That Matter, Hon
Performance Appraisal: A Guide to Greater Productivity, Olson
Planning For Organizational Success, Kaufman
Quantitative Management, Schneider
Quick Legal Terminology, Volkell
Quick Medical Terminology, Smith
Quick Typing, Grossman
Quickhand, Grossman
Skills for Effective Communication: A Guide to Building Relationships, Becvar
Speedreading for Executives and Managers, Fink
Successful Supervision, Tagliaferri
Successful Time Management, Ferner
Using Graphs and Tables, Selby
Using Programmable Calculators for Business, Hohenstein

Look for these and other Wiley Self-Teaching Guides at your favorite bookstore!

*In preparation.

MINERALS, ROCKS, AND FOSSILS

R. V. DIETRICH and **REED WICANDER**

Central Michigan University

A Wiley Press Book

JOHN WILEY & SONS, INC.

New York · Chichester · Brisbane · Toronto · Singapore

Publisher: Judy V. Wilson
Editor: Alicia Conklin
Managing Editor: Maria Colligan
Composition and Make-up: Cobb & Dunlop Publisher Services, Inc.

Library of Congress Cataloging in Publication Data

Dietrich, Richard Vincent, 1924–
    Minerals, rocks, and fossils.

    (Wiley self-teaching guides)
    Includes index.
    1. Minerology. 2. Petrology. 3. Paleontology.
I. Wicander, E. Reed, 1946–   . II. Title.
QE363.2.D53 1982   552   82–20220
ISBN 0–471–89883–X

Printed in the United States of America

83 84 10 9 8 7 6 5 4 3 2 1

Contents

Acknowledgments

Several of our illustrations and ways of expressing ideas have been used in previously published pamphlets and books by the senior author. Especially noteworthy are the following publications: *Virginia Mineral and Rocks,* fourth edition (Virginia Polytechnic Institute); *Mineral Tables: Hand-Specimen Properties of 1500 Minerals* (McGraw-Hill Book Co.); *Geology and Virginia* (University Press of Virginia); *Rocks and Rock Minerals*—coauthored with B. J. Skinner (John Wiley & Sons); *Stones: Their Collection, Identification, and Uses* (W. H. Freeman & Co.); and *Mineralogy: Concepts, Descriptions, Determinations,* second edition—with L. G. Berry and Brian Mason (W. H. Freeman & Co.). All publishers (shown in parentheses) graciously gave permission for us to borrow freely from the listed publications.

We acknowledge the artistry of Greg Kubczak and John Rowe, who prepared most of the illustrations, and of Greg Swartz, who sketched Figures 1–4 and 2–1; also the patience of Janet Solocha, who helped compile the Mineral Determinative Tables given in Appendix II.

We wish to thank Richard A. Bideaux of Tucson, Arizona; Professor Cornelius Hurlbut, Jr., of Harvard University; and Professor Wayne Moore of Central Michigan University for reviewing the manuscript; and Karl Weber, who acted as developmental editor.

R. V. Dietrich

Reed Wicander

CHAPTER 1

Getting Started

Minerals, rocks, and fossils are interesting for many reasons—and every collector has his or her own special reason for collecting them.

One might, for instance, collect minerals, rocks, and fossils simply for their beauty of color or form. Nearly all fine jewelry contains some mineral or group of minerals. Some collectors cut and polish the minerals and rocks that they collect—an enjoyable and rewarding pastime.

Others collect minerals, rocks, or fossils just for the joy of collecting something. These hobbyists have the added satisfaction of enjoying the great outdoors while building up a catalog of all of the various specimens found in their favorite areas.

Still others, primarily geologists, know that the history of the earth is contained in its minerals, rocks, and fossils, and that learning about them is absolutely essential to unraveling that history. Rocks and particularly the fossils they contain may tell what an area was like in the past—its landscape, its climate, and even the kinds of plants and animals that lived there. Rocks, truly pages from the history book of the earth, give a fascinating picture of time for the hobbyist as well as the expert.

The list of why people collect could go on and on. But whatever your particular reason for collecting, once you begin, you will find it hard to stop. You will probably make it a lifetime pursuit, as you learn more and more about the objects you collect.

As a beginning, read this book. You will learn such basic things as how to identify and name most of the common minerals, rocks, and fossils; how minerals, rocks, and fossils are formed; and where they occur.

REVIEW QUESTIONS

Give three reasons why people collect minerals, rocks, and fossils.

1. _____
2. _____
3. _____

What are some of the things minerals, rocks, and fossils can tell us about the history of the earth?

_____

_____

Why are *you* interested in collecting and learning about minerals, rocks, and fossils?

_____

_____

_____

_____

WHERE TO COLLECT MINERALS, ROCKS, AND FOSSILS

Minerals, rocks, and fossils can be found almost anywhere. The local geology of your area and the accessibility of collecting sites, however, will determine what types of minerals, rocks, and fossils you can expect to collect near where you live. And, as with any collecting hobby, your road to success is in knowing what to collect, where to collect, and how to collect.

One of the best ways to find out what collecting opportunities are available in your area is to attend a meeting of a local mineral, rock, and/or fossil club. You will meet people there who will be eager to share their knowledge with you and to help you find what is available within your area. In addition, many clubs organize field trips to other areas where you will be able to collect different minerals, rocks, or fossils. If there isn't a club nearby, visit a local museum, high school, or college. Someone there will probably be happy to help you get started.

EQUIPMENT YOU NEED TO GET STARTED

The equipment you will need to collect minerals, rocks, and fossils is minimal. If you do not already own it, you will find it fairly inexpensive to buy (Figure 1–1).

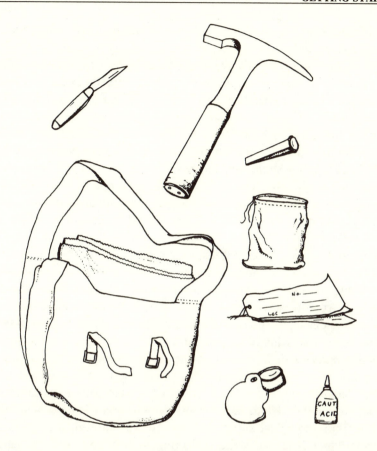

Figure 1–1. Equipment needed for collecting minerals, rocks, and fossils. Clockwise from upper left: knife, geological pick, chisel, sample bag, labels, dilute hydrochloric acid, hand lens, collecting bag, and paper.

The first thing you should have is a good *geological pick*, although a bricklayer's hammer will do. There are two kinds of geological hammers. Which one you should choose depends largely on the type of collecting you plan to do. One is a pick type with a sharp point on one end; the other has a wide chisel on one end. Both have a flat, square head on the other end, which is used for removing rocks from an outcrop, for breaking up rocks, and for shaping them. On the pick type, the pointed end is used for chipping out minerals or fossils on small pieces of rock. On the other type, the chisel end is used primarily for digging in soft sediments. For most collectors, the pick type of hammer is by far the more useful and more practical of the two. It can also be used for digging in soft sediments and thus is more versatile than the chisel type. Geological hammers can be purchased at many large hardware stores and can be ordered from nearly all geological or prospectors' supply dealers. Before you use your pick, be sure that you also purchase and use protective eye goggles, as we caution you in the following section.

A *hand lens* helps the collector to identify small minerals, small rock constituents, and microfossils. Many of the features necessary for mineral identification, such as cleavage and crystal shape, would be difficult, if not impossible, to see without a hand lens. And, as one needs to know the mineral makeup of most rocks in order to identify them, a hand lens is a necessity for those in which the constituent minerals are small. Also, even in large fossils, many delicate features cannot be seen without the aid of a hand lens. A 10-power (10×) hand lens is the most satisfactory for most macroscopic observations. They can be purchased from many stationery stores, hobby shops, and hardware stores, as well as from geological or prospectors' supply houses. The hand lens can be carried in one's pocket. It also can be hung from a chain or a piece of rawhide and worn around the neck—this is preferred by many collectors because the lens is readily accessible and less likely to be lost.

A *collecting bag,* or *knapsack* is also essential. It will enable you to carry more than will fit in your pockets, and when you are collecting far from your car or other means of transportation, it will save you the trouble of making trips back and forth as you collect. Appropriate collecting bags can be purchased at most sporting goods stores or geological supply houses. Be sure to buy one that is comfortable to carry and not so large, that when full, you can't carry it because it is too heavy!

Wrapping material such as tissue or newspaper should be taken along on all collecting trips so you can wrap, and thus protect, any well-preserved or delicate specimens. *Sample bags,* either cloth or paper, should be used to separate specimens from one another. Cloth sample bags, which come in different sizes and can be ordered from geological supply houses, are very useful; they have sewn-on labels upon which you can write collecting information, and drawstrings for closing. Most of the time, however, ordinary paper sacks will suffice, and you can write directly on them. In either case, you should use a permanent marking pen for recording the information.

A *notebook* is also essential. It should be spiral-bound or sturdily sewn, and small enough to be carried easily. Get into the habit of using a notebook to record all information about a collecting trip while you are at the collecting site, rather than relying on your memory. Information about a trip should include the date, the geographic location, the type of outcrop, the kind of rock, its formational name and geological age, the types of material collected, and any other information you think pertinent. You also may add a field sketch or a description of the site.

Other useful, but not always essential, items include:

1. *Mineral, rock, or fossil guidebooks.* Depending on the area and your knowledge of it, these may or may not be necessary. If you have been to the area before or are knowledgeable about the material you will be collecting, you probably won't need a guidebook. On the other hand, if the area is new to you, a guidebook may be extremely helpful.

2. *Topographic and/or geological maps.* Again, depending on your knowledge of the area, you may or may not want to have such maps with you. See Appendix I for information on getting and using these maps.
3. *Masking tape.* This is useful for sealing paper bags, for holding large fossils or rock specimens together, and even for wrapping some small specimens. Among its other advantages, it can be written on easily.
4. *Trenching shovel.* This implement is especially useful in an area of soft, generally unconsolidated sediments or highly weathered sedimentary rocks.
5. *Chisel and/or sledgehammer.* A chisel is useful for chipping specimens out of a rock; a sledgehammer, for breaking up very large rocks.
6. *Camera.* A camera is almost a necessity for recording what some collecting sites look like. Slides or photographs are very helpful in reminding you about a site, especially if you haven't been there in a long time, and also for showing to other collectors.
7. *Dilute hydrochloric acid (10 percent HCl, or "muriatic acid," to 90 percent water).* This is especially useful in determining whether calcite or dolomite is present and thus in identifying a rock as a limestone rather than a dolostone; or vice-versa.

Some other items you should have are discussed in the following section.

DO'S AND DON'TS FOR THE COLLECTOR

Before you collect minerals, rocks, or fossils, you should know where to look and have an idea of what you want to collect. Thus any field trip begins at home with your initial decisions and preparations.

One of the best ways to learn about collecting is to go on a trip organized by an experienced collector. Such trips are arranged by local clubs, schools, and museums; you can find out about them by contacting the appropriate groups.

In collecting, as in other activities, common sense is the essential rule. In addition, respect for nature and for the rights of others is important. Unfortunately this is not always recognized, and the thoughtless actions of a few can spoil otherwise enjoyable occasions for everyone.

Although mineral, rock, or fossil collecting is a relatively safe pastime, certain precautions must be taken while in the field (Figure 1–2). You should always be alert to potential dangers at a collecting site. No specimen is worth the price of an injury or a potentially fatal accident. A *hard hat* should be worn wherever there is any possibility of falling rock. Do not work directly below others on a slope above you; any rocks they dislodge might fall on you, or they themselves might even fall on you. Be particularly alert in quarries where there are overhangs. Many overhanging rocks are already shattered or fractured, ready to fall at any time. Even a hard hat will do little to protect you from an overhanging ledge that collapses.

Figure 1–2. Collecting at the outcrop. Note that these collectors have ignored the common sense rule of wearing a hard hat. (Sketch from "Geological Excursions in Southwestern Virginia," published by the Department of Geological Sciences, Virginia Polytechnic Institute.)

Hard-toe boots with nonslip soles and *eye goggles* should also be used on field trips. Many a foot has been spared a bruise or break by the wearing of sturdy hard-toe boots. And, whenever you are breaking up a rock or chipping out a sample, the wearing of shatterproof eye goggles is mandatory. It is only too easy to chip out a small grain of something and have it hit your eye before you can blink, causing an injury if you are lucky, or blindness if you are not. Therefore, before you do any hammering on an outcrop or try to dislodge a specimen, put on your safety goggles. The few seconds it takes is a small price to pay to protect your eyesight.

Other hazards of collecting include the usual outdoor nuisances: insects, poisonous plants, and animals. Learn to identify and avoid, in particular, poison ivy, poisonous snakes, scorpions, ticks, and spiders.

While the specific techniques employed in mineral, rock, or fossil collecting may vary slightly, depending on what you are collecting, there is a general code of etiquette that everyone should follow. It includes the following:

1. First and foremost, obtain the permission of the landowner before entering or collecting on private property. Most landowners will consent to collecting on their property if they have been asked first and told the reason for it. Unfortunately, many collecting sites have been declared off-limits because of the thoughtlessness of a few individuals who trespassed on private property without first asking permission.

2. Collect only sufficient material for your needs, and be as neat as possible at the collecting site. This means filling in holes after collecting so that livestock won't fall in or stumble and injure themselves, and taking your litter with you when you leave.

3. Never blast without the express permission of the landowner, and even then only under rigid safety precautions.

4. Leave machinery, supplies, and equipment where they are.

5. In farm areas, do not trample on or otherwise damage growing crops.

6. Do not disturb livestock, and be sure to leave gates open that are open and to close any gates you find closed.

7. Always remember that you are a guest and that your actions reflect on collectors in general. How you behave may determine whether future groups will be allowed to collect at the site.

One can collect on some public lands, but not on others. In any case, it is a good idea to check local and state laws that may be relevant. Some states have laws that apply to collecting on state lands; you should be familiar with these laws before collecting there. The antiquities law of 1906 (U.S. Code Section 34, Statute L-255) governs the collecting of antiquities on federally owned land. In the past, this generally has not applied to the collection of minerals, rocks, and invertebrate fossils; however, as the law is currently undergoing reinterpretation, it would be wise to check its application before collecting on federal property. Circular 2147 of the Bureau of Land Management, U.S. Department of the Interior, covers the laws and regulations that affect mineral and fossil collecting on lands under its jurisdiction.

Finally, although road cuts make excellent exposures for collecting, roadside collecting is generally prohibited by state police, particularly along interstate highways. Even if you park in a designated rest area or on another road, you should check with the police before doing any collecting in highway right-of-way areas.

REVIEW QUESTIONS

What should you do before you try to collect minerals, rocks, or fossils on any property, either privately or publicly owned?

_____

_____

_____

If you have never collected before, what is the best way to get started?

_____

_____

_____

What is the basic equipment needed for collecting? How is each piece of equipment used?

_____

_____

_____

_____

_____

_____

List three hazards of collecting.

1. _____

2. _____

3. _____

What should you do before a collecting trip to help assure a successful outing?

_____

_____

What are the basic rules of etiquette in collecting?

_____

_____

_____

_____

CLEANING AND PREPARING MINERAL, ROCK, AND FOSSIL SPECIMENS

Finding your minerals, rocks, or fossils is only part of making a collection. Many specimens collected in the field need to be cleaned and prepared for proper labeling and display. Minerals, rocks, and fossils each involve special cleaning and preparation techniques; we shall cover only widely used techniques that are generally applicable to all three. To learn more about special techniques, consult one of the books devoted exclusively to the subject, such as Kummel and Raup (1955), listed at the end of Chapter 4.

While in the field, the best advice is to be careful to wrap your specimens suitably so they will arrive home in good condition.

When collecting minerals, take great care to prevent damage to them. In the field, wrap the mineral specimens in newspaper or tissue paper and masking tape. Wrap small, fragile specimens in cotton and store them in, for example, plastic vials.

Soap and water combined with a brushing will clean most mineral specimens rather well. One should not coat mineral specimens with shellac or other coatings; this detracts from their natural beauty.

Some minerals can be collected as loose specimens; others are integral parts of rock materials and thus must be chipped out. Fortunately, many minerals are best displayed when still attached to their associated minerals or rocks. If a specimen free from its host rock is desired, great care must be taken when trying to remove it so as not to hit, and accidentally break, it. The judicious use of your rock hammer and chisel and, in some cases, such things as dental tools, along with lots of practice, will determine the degree of your success.

Although most rocks are easier to clean and prepare than minerals or fossils, a certain talent is required for shaping them the way most collectors do. Rocks can either be collected as loose specimens or hammered or chiseled off an outcrop. For uniformity in your collection and easier storage, most of your rock specimens should be about the same size and shape.

Shaping can be achieved by striking the rock along its edges to break off the sharp pieces and gradually remove any protrusions. By doing this systematically, you can shape your rock into a rectangle and, by repeated blows to its edges, into the final size and shape you desire (Figure 1–3). This technique takes practice, but is not really difficult to master, and the results are well worth it.

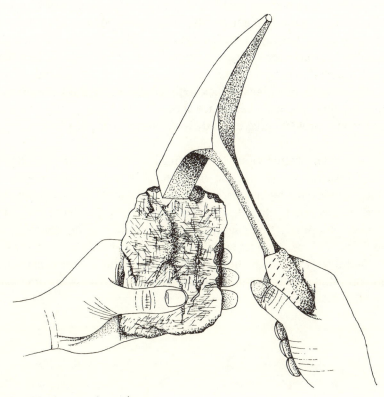

Figure 1–3. Shaping a rock specimen.

If a rock specimen does not crumble when immersed in water, a good soap and water bath plus brushing will usually be sufficient to clean it. If the rock is friable (that is, crumbles easily) or tends to disintegrate in water, a light brushing to remove loose dirt will have to suffice. In most instances, cleaning should be unnecessary because you will collect freshly broken pieces.

A common coating on some rocks is rust, an iron oxide. Rust can be removed from some rocks, at least in part, by brushing or other physical means. For other specimens, however, it may be necessary to employ chemical methods, such as soaking the specimen in oxalic acid.

Many of the fossils you collect will already have been weathered out of their containing rock and thus need only to be picked up. They can be wrapped in newspaper, or, if they are abundant and appear sturdy, simply placed in your sample bag. If, however, the fossils are fragile, wrap them in newspaper, or, if they are small, wrap some masking tape around them. If there is a chance that the fossil will break because it is extremely delicate, you can coat the specimen with shellac, varnish, or some plastic hardening compound. In such cases, it is a good idea to leave some of the matrix (host or containing rock) underneath the specimen to act as a support. After the protective coat has dried, wrap the specimen in paper.

Some fossils, particularly plant fossils, occur as carbonaceous (black, carbon-rich) impressions on the layers (often called bedding planes) of a rock. Such specimens should be covered with a thin protective layer of, for example, shellac or varnish and wrapped in paper. In well-bedded rocks, plant fossils and other fossils are generally found by breaking the rock so that it splits along its bedding planes. If a plant impression is present, the surface should be blown or gently brushed clean before applying a protective layer of shellac, varnish, or plastic hardening compound.

You should not try to free fossils that are firmly embedded in a rock while you are in the field. Instead, you should use either the pick end of your hammer or your hammer and a chisel to chip away much of the rock around the specimen, and then bring back the specimen with any still-attached rock. (Remember to wear your eye goggles when chipping.) You can then clean or try to remove the fossil at home under better conditions and with special tools such as needles and probes used by dentists (Figure 1–4). If the specimen should break while you are trimming it in the field and you still want it, pack up the broken parts and take them home where you can glue them back together.

Figure 1–4. Cleaning a fossil specimen.

Sometimes the original shell material has been removed, leaving only a mold of the specimen. A cast of the specimen can be made by cleaning out the mold, coating it with a thin layer of oil, and filling in the mold with plaster of Paris. After the plaster has hardened, you can carefully break away the surrounding matrix until the cast is free.

Frequently, fossils are in soft rocks, such as shales or marls, or in rock materials that are not truly solid—that is, poorly consolidated rocks. Rather than to try to remove such fossils from the rock in the field, it may be easier to bring the material home and clean it there.

A good rule is to clean and prepare your fossils at home where you can do it slowly, using proper tools. The type of cleaning and preparation you use will depend on the fossil itself—how fragile it is, what type of matrix it is embedded in, what features you want to highlight, and perhaps other considerations.

The simplest cleaning procedure is to let the specimen soak overnight in soap and water. This will remove, for example, loose clay adhering to the specimen. Unless the specimen is extremely fragile, it can be cleaned further by giving it a good brushing with a toothbrush. For most specimens, however, more work will be required, particularly to remove the matrix from around the fossil. You will generally need a small hammer, a set of chisels, long needles, and probes such as those dentists use (Figure 1–4). Most of these can be purchased at a hardware or hobby store. You also may want a large magnifying glass, tweezers, and even a pair of long-nosed pliers for removing excess matrix. The proper "touch" can be developed only with practice; it is advisable first to practice on specimens of which you have several so that you can afford to make mistakes. However, if a specimen is broken during cleaning, in some cases, it can be glued together with any of the common household glues or cements on the market.

Some fossils can be removed or cleaned by using a dilute acid. For example, if the specimen has been replaced by a noncarbonate mineral and is embedded in limestone, you can dissolve the matrix away by immersing the specimen in dilute hydrochloric or acetic acid. You also may be able to remove carbonate overgrowths or matrix from a specimen by selectively applying a drop of dilute acid with an eye dropper and, when it reaches the fossil, quickly washing it off.

The techniques just discussed apply to the cleaning and preparation of *megafossils,* which are fossils large enough to be seen with the naked eye. These are what you are most likely to collect, at least at first.

Microfossils are fossils that have to be studied under the microscope. The two common microfossils in which you might be interested are foraminifers and ostracods; these are described in Chapter 4. The usual preparation of these fossils involves these steps:

1. Break the rock into pieces about 1/2 inch in diameter, and place the pieces in an oven at about 300°F, until they are thoroughly dried.
2. Place the dried pieces in a jar and add kerosene to cover them completely.
3. Allow the mixture to sit overnight.
4. Pour off the kerosene and add water; the sample will then begin to disintegrate.
5. Allow the sample to disintegrate overnight.
6. Wash the remaining material through a 200 mesh screen. The microfossils will be left on top of the screen; after drying, they can be stored in a vial or an envelope until they are picked and mounted on special 1-inch by 3-inch slides that have been coated with a water-soluble glue (Figure 1–5).

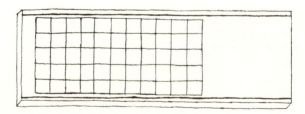

Figure 1–5.   Typical microfossil slide with 60 squares, each square containing the same species.

In this section, you have learned how to collect, clean, and prepare minerals, rocks, and fossils. Now you should review these suggestions before trying to apply what you have learned.

REVIEW QUESTIONS

How should fossils be protected to get them safely back from the field?

_____

_____

If a fossil is firmly embedded in a rock, how should you collect it?

_____

_____

How should you clean and prepare such a fossil at home?

_____

_____

What is the best way to clean almost any mineral or fossil?

_____

_____

How can you free minerals and fossils that are embedded in limestone?

_____

_____

How can you collect dinosaur footprints? (*Hint.* Can footprints be considered molds?)

_____

_____

_____

How should you prepare your rocks so they could be used in a display?

_____

_____

_____

What tools and equipment should you have for cleaning and preparing fossils?

_____

_____

_____

What is the best way to collect and clean fossil plants?

_____

_____

_____

How does cleaning and preparing microfossils differ from cleaning and collecting megafossils?

_____

_____

_____

_____

LABELING, STORING, AND DISPLAYING YOUR COLLECTION

Without labels, most minerals, rocks, and fossils are little more than interesting or curious objects. The serious collector will want to have all specimens labeled properly. Many collectors even number their specimens and keep a catalog that contains a permanent record of each specimen.

There are two kinds of labels. The first kind is the one attached to the specimen when it is collected in the field. As already stated, each and every specimen collected, whether it is a mineral, rock, or fossil, should have a label that describes where it was collected. The locality should be described as exactly as possible; the label should include both geographic information and whether the specimen was collected from a rock exposure, such as a river bank or quarry face, or from loose material, either near or far from its probable parent exposure. If there is more than one formation exposed at the site, the particular formation the specimen was collected from should be noted. The scientific name of the specimen, if you know it, can also be put on the field label.

Some collectors only number their specimens in the field and record the specimen number and information about it in a notebook. We think that it is a good idea to put more than just a number on field labels, in case you lose our field notebook. In this way you can be sure not to be left with a collection of numbered specimens and no specific information as to where they came from.

The second kind of label is the permanent label you attach to the specimen when you put it into your collection. It should be neatly written or typed and should contain the following information:

1. Name of the specimen
2. The exact geographic locality where it was collected
3. Name and age of the containing rock
4. The collector's name and the date it was collected
5. A specimen number referenced to your catalog, if you keep one.

In making a catalog, every specimen is given a number, and these numbers are entered sequentially in the catalog. Many collectors paint a number on the specimen, so that the specimen can still be identified if the label is lost. Each numbered entry has at least the same information as the label for the specimen. In fact, many collectors use their catalogs to go into more detail about each specimen than can be put on a label; for example, they may include photographs of the collecting site.

Also, the catalog can be cross-referenced with, for instance, one set of entries by catalog number and another by some type of classification scheme.

Figure 1–6 shows a sample label for a fossil trilobite, along with an 1824 label for a gold specimen from the collection of Colonel Washington A. Roebling. Roebling's collection, which included more than 16,000 mineral specimens—at least one specimen of all but 12 of the named minerals from the index of Dana's *System of Mineralogy* (sixth edition, 1892), is now in the Smithsonian Institute, Washington, D.C.

Specimen No.	NMNH 8462

Name: *Cryptolithus tesselatus*

Geologic Formation and Age: Martinsburg (Ordovician)

Geographic Locality: Swatara Gap, N.W. of Lebanon, Pennsylvania

Collector: G.A. Cooper

Date: July 4, 1956

A

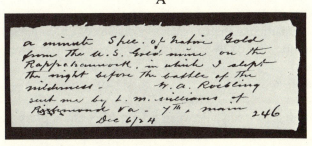

B

Figure 1–6. (A) Sample label for the trilobite *Cryptolithus tesselatus,* showing geographic locality, geologic formation and age, collector, and specimen number. (B) Label for gold specimen from collection of Colonel Washington A. Roebling's, which is now in the Smithsonian Institution. (Photographs courtesy of Smithsonian Institution.)

The index of your catalog can be set up on any basis. If you are a mineral collector, you may want to arrange your entries by the chemical classification used by many professionals; on the other hand, you may prefer to arrange the entries alphabetically by mineral name, or according to some other system. A rock collection might be classified by genetic class—for example, igneous, sedimentary, or metamorphic. A fossil collection might well be classified by phylum. In any case, the specimen's name and its number need to be recorded so that you can go to the number in the catalog and find out where the specimen was collected and anything else you have recorded about it.

Storing a collection can range from keeping a box of specimens in a drawer or closet to housing them in permanent storage cases, such as some of the commercially available ones, built especially for such purposes. The type of storage you choose will depend on how delicate your specimens are, how valuable they are, and how much money you want to spend. However you decide to store your collection, you should make sure that the specimens will be safe from breakage and other damage.

How you display your collection will depend on the type of collection you have and the amount of time, energy, and money you want to spend in making a display. The easiest way to display your collection is to keep each specimen and its label in its own tray or box. This method has the advantage of both simplicity and versatility. The boxes can be arranged any way you want in drawers or on shelves and, if you want to change the emphasis of your collection, you have only to rearrange your boxes.

Some collectors, especially those who collect fossils and minerals, may want to display their specimens in *Riker Mounts*. These are shallow glass- or plastic-topped cardboard boxes filled with cotton. Very delicate or fragile specimens, in particular, are frequently displayed this way.

Another method is to construct a display board to which the specimens are attached—in some cases, together with pictures or drawings. Such boards are usually used for a particular theme, such as a display of different types of volcanic rocks or an evolution-based display of fossils.

REVIEW QUESTIONS

What is the difference, if any, between a field label and a permanent label?

_____

_____

_____

What is the least information you should include on a field label?

_____

_____

_____

What is the least information you should include on a permanent label?

_____

_____

Why is it a bad idea to put only a number on your specimen and then record your field information in a notebook?

_____

_____

What are the advantages and disadvantages of making a catalog for your collection?

_____

_____

What are three ways you might catalog your collection?

    1. _____

    2. _____

    3. _____

What are three ways you might display your collection?

    1. _____

    2. _____

    3. _____

Why is it important to store a collection properly?

_____

_____

How do you plan to store your specimens?

_____

_____

_____

JOINING A CLUB

One of the many advantages of collecting minerals, rocks, and/or fossils is that it offers you the opportunity to meet and socialize with other collectors while you learn about your specimens.

Wherever there are people who collect them, there is likely to be a club devoted to minerals, rocks, or fossils, or to all three. These clubs are dedicated to communicating the joy of collecting to their members as well as to helping members increase their knowledge about the objects they collect. Even though most clubs are organized by amateur collectors, some are affiliated with schools or museums.

Most clubs have monthly meetings at which club business is discussed and some type of program, generally educational, is presented. Most of the presentations are talks by a professional or another amateur about some aspect of collecting or some other topic dealing with minerals, rocks, or fossils.

Clubs also organize field trips to collecting sites, and are often able to get into quarries or on to private property that is off-limits to individuals. Besides the camaraderie that frequently comes from sharing a common interest, the opportunity to participate in organized field trips is a real advantage of joining a club.

To find a club in your area, first try the telephone book. If no club is listed, then check with science teachers at the local schools or with museum personnel. Usually they will know if there are any clubs in the area, or who the local collectors are and how they may be contacted.

Joining a club can add immeasurably to your enjoyment and knowledge of minerals, rocks, and fossils. Club members come from all walks of life, sharing a common bond in their interest in collecting. They are proud to be called "rockhounds" (involved youngsters are often called "pebble pups"). Most members not only are willing, but eager, to share their knowledge with newcomers; they need only to be asked. Become involved!

WHERE TO FIND HELP

You will be learning how to identify minerals, rocks, and fossils in the next three chapters. Nonetheless, whether you are just starting or have been collecting for years, you will sometimes find a specimen whose identity completely baffles you. When this happens, you should turn to a more experienced collector or, in some cases, to a professional.

If you live in a highly populated area, there will probably be a nearby museum or nature center, and perhaps several schools in which geology is taught. Most geology or earth science departments at colleges and universities will have someone who either knows what you have discovered or can direct you to a specialist or a book that will help you. Museums and nature centers usually have personnel who can help you in your identification. Many of these professionals also are collectors and are only too glad to help a fellow collector. They also are likely to be familiar with the area and what is available there, and thus can tell you about other localities where you may collect. When you want to talk to one of these experts, it is wise to call ahead to make an appointment; you may be unable to see the person if you drop in unannounced. Also, in some cases, you can find out what you want to know over the telephone.

VISITING MUSEUMS

Valuable information and ideas often come from visiting museums. Nearly all large cities have museums, and most of these museums have a section devoted to minerals, rocks, and/or fossils. Many universities and colleges also have geological museums. There you will find the finest specimens on display and perhaps have a chance to see many minerals, rocks, and/or fossils that you have only seen illustrated or have read about in books.

It is well worthwhile to visit museum displays whenever possible. Not only are they informative, but you may get some new ideas about how to display your own specimens.

TRADING, BUYING, AND SELLING SPECIMENS

After you have been collecting for a while, you will accumulate many specimens that you no longer want or need. At this point, you should consider selling or trading some of your unwanted specimens. How do you go about it?

Many mineral, rock, and fossil clubs have shows where collectors can trade, buy, or sell specimens. This is usually one of the functions of a club.

Trading specimens is a good way to upgrade your collection or to fill in gaps. You should take only your best duplicate specimens to shows where you plan to trade. You should have as much information as is known about each specimen on its label. A good trader will look over what is available with an eye for what will improve his or her collection and then try to trade value for value. You must remember that, to get a good specimen, you will have to have a good specimen to give up.

Many times, you will not be able to get what you want by trading. Then, you may want to consider buying a specimen or specimens in order to improve or complete your collection. As with any purchase, let the buyer beware. Look at as many different displays as possible and try to get a "feel" for the worth of the specimen you are buying. If you can't get to a show but must buy from a dealer, buy only from a reputable one. Ask around about different dealers' reputations.

If you have some rare or very well-preserved specimens, you might consider selling them. Club shows and ads in magazines directed to mineral, rock, or fossil collectors are good ways to sell your material. Large dealers or suppliers may buy some of your specimens. You should query them as to their needs before sending any specimens. You are likely to get more per specimen, however, from other individual collectors.

Just as identifying and displaying your specimens are enjoyable parts of collecting, so too can trading, buying, or selling them add to your overall enjoyment.

REVIEW QUESTIONS

What are the purposes of a mineral, rock, or fossil club?

_____

_____

_____

_____

_____

How do you go about finding a club in your area?

_____

_____

_____

What are two sources of help in identifying difficult specimens?

_____

_____

_____

How can museums and their personnel help you in identifying your specimens?
How else are museums useful to you?

_____

_____

_____

_____

For what purpose would you want to buy, sell, or trade specimens, and how and
where would you go about it?

_____

_____

_____

_____

_____

BOOKS AND MAGAZINES FOR THE AMATEUR

A wealth of information is available about minerals, rocks, and fossils for the individual who wants to learn more about them and to go into more detail than this book is intended to provide. The following periodicals contain additional information for the collector. In addition, several state and federal geological surveys also publish books and pamphlets dealing with minerals, rocks, and fossils. We also list some good reference books at the end of each of the following chapters.

Periodicals:

> *Lapidary Journal*. Lapidary Journal Inc., Box 80937, San Diego, Calif. 92138.

> *Natural History*. American Museum of Natural History, Central Park West at 79th Street, New York, N.Y. 10024.

> *Rock and Gem*. Behn-Miller Publishers Inc., 16001 Ventura Boulevard, Encino, Calif. 93436.

> *Rocks and Minerals*. Heldref Publications, 4000 Allemarle Street, N.W., Washington, D.C. 20016.

> *Rockhounds*. Latham Publications Inc., Box 328, Conroe, Texas 77301.

Other Sources:

> American Federation of Mineralogical Societies, 4139 South Van Ness Street, Los Angeles, Calif. 90062. This is the national headquarters. There are regional federations, and many local, affiliated groups, some of which issue official newsletters and similar publications.

CHAPTER 2

Minerals

Minerals are extremely important economically, aesthetically, and scientifically. Economically, minerals are the source of many of the things we use and enjoy in our current way of life. Aesthetically, minerals enrich our lives as gems, and as specimens in their natural surroundings, in collections, or in museum displays. Scientifically, minerals make up the data bank from which we learn about geological processes and how, for example, to synthesize materials in short supply.

For ages, the word *mineral* and related words were used to refer to anything not animal or vegetable. More recently, at least so far as the mineralogist is concerned, *mineral* has taken on a much more precise meaning.

In this chapter, you first learn the definition of *mineral* and examine the diverse aspects of that definition. Then you learn how minerals are named and classified, how they are formed, where they occur, how they are described and identified, and what mineral-like substances are termed *mineraloids, glasses,* or *macerals.*

REVIEW QUESTIONS

Can you think of and list some of the ways minerals are important in human activities?_____

_____

_____

(Did you include some of these—as a constituent of soil; as the raw material for metals; as raw materials for building materials such as cement, concrete, plaster, and bricks; as a source for many chemicals?)

Of the topics to be covered in this chapter, which do you expect to enjoy the most?

_____

(Check this answer after you finish the chapter to see if you have changed your mind.)

DEFINITION OF MINERAL

A mineral is a natural solid, generally formed by inorganic processes, with an ordered internal arrangement of atoms and a chemical composition and physical properties that either are fixed or vary within a definite range.

There are five chief aspects to this definition. A mineral must:

1. Be natural
2. Be solid
3. Have an ordered arrangement of its constituent atoms
4. Have a chemical composition that is fixed or varies within a definite range
5. Have physical properties that are fixed or vary within a definite range.

In addition, the definition states that minerals are, in general, formed as a result of inorganic processes.

Each of these requirements for a mineral warrants further consideration.

That a mineral must be *natural* simply means that a mineral must have been formed in nature. Man-made materials, even those designated by such names as "mineral spirits" or "mineral oil," are not minerals in any sense. On the other hand, many gemstones, that are manufactured but fit the definition in every other way, are accepted by both mineralogists and the legal profession under the appropriate *synthetic mineral* designation. (The term *simulated mineral* has a different meaning; it is the name applied to any material that resembles and is used to substitute for a mineral or its synthetic equivalent. For example, ruby or synthetic ruby may be simulated by such diverse red substances as natural or synthetic spinel, glass, or even plastic.)

The requirement that a mineral be *solid* eliminates natural fluids such as petroleum and natural gas from consideration as minerals. Many mineralogists, however, accept native mercury, a liquid at normal temperatures and pressures, as a mineral species.

The *ordered arrangement of constituent atoms* is, in many ways, the most critical part of the definition. The arrangements, which are three-dimensional arrays of atoms such as the one shown in Figure 2–1, are called *crystal structures*.

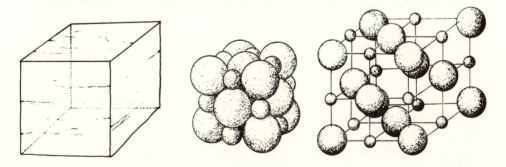

Figure 2–1. The cubic cleavage fragment (left) and the internal structure of halite, common table salt. The lattice diagram (right) shows the relative positions of sodium and chlorine; the relative sizes and packing arrangement are depicted in the center.

They may consist of atoms or ions of only one chemical element or of combinations of two or more elements. The arrangements depend upon such things as the sizes of the constituent atoms or ions and the way they are bonded together. Each mineral has its own unique arrangement of its constituent elements. Under favorable conditions of formation, the ordered arrangements may be expressed by external crystal forms (Figure 2–2).

Figure 2–2.   Quartz crystals ("rock crystals") from Rich Valley, Smyth County, Virginia. (Photograph by G. K. McCauley.)

The fact that both the *chemical composition and physical properties may either be fixed or vary within some definite range* has a number of interesting aspects. A mineral's chemical composition is generally expressed as a formula that gives the proportions of its constituent elements (Appendix VI). For example, if silicon (Si) and oxygen (O) atoms are present in the ratio of 1:2, the composition can be represented by the formula SiO_2, which means that for each silicon atom present there are two oxygen atoms. This is quartz (and other SiO_2 minerals), an example of a mineral with a fixed chemical composition.

As long as its crystal structure is set, a mineral with a fixed chemical composition also has fixed physical properties. This crystal structure restriction must be imposed because some elements and groups of elements may be combined in different ways, each having its own physical properties—(for example, diamond and graphite are both pure carbon.) The name usually applied to the phenomenon whereby the same chemical substance may have different crystal structures is *polymorphism,* which means "many forms."

As the definition for mineral indicates, both the chemical composition and the physical properties of a mineral may vary, only so long as they vary within a definite range. Such minerals constitute series widely referred to as *solid solution series*. The minerals of such series have physical properties, such as specific gravity, that vary with the variation of chemical composition (Figure 2–3). A well-known example of a solid solution series is the plagioclase feldspars (see Appendix II), which range from albite ($NaAlSi_3O_8$) to anorthite ($CaAlSi_2O_8$).

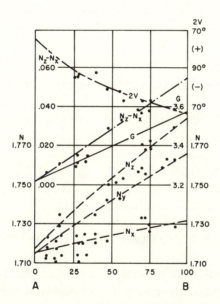

Figure 2–3. A hypothetical solid solution series. For example, *A* could be a sodium-aluminum silicate and *B* its equivalent sodium-iron silicate. On the graph, *G* is specific gravity; *Nx, Ny, and Nz* are indices of refraction (that is, the velocity of light transmission as compared with that through air); and *2V* is a measurable optical property. (Modified after A.N. Winchell and Horace Winchell, *Elements of Optical Mineralogy*, 4th ed., Part 2. © 1951, John Wiley & Sons, Inc., New York.

The final aspect of the definition, *generally formed by inorganic processes,* is considered superfluous by many mineralogists. It is a throwback to the view that even inorganic substances produced by plants and animals (for example, the aragonite that constitutes pearls) are not minerals *per se.* This view though not completely abandoned, is no longer prevalent. Also, this restriction never was interpreted to preclude all organic compounds from mineral status; a number of solid hydrocarbons, calcium oxalate, and so on, have long been recognized as minerals.

REVIEW QUESTIONS

What are the five chief aspects of the definition of a mineral?

1. _____
2. _____
3. _____
4. _____
5. _____

What is the main difference between synthetic sapphires and simulated sapphires?
_____

What is polymorphism? Give an example. _____
_____
_____

What is a solid solution series? _____
_____
_____

THE NAMING AND CLASSIFICATION OF MINERALS

The names of some minerals are so old that their origins are lost in antiquity. In the latter part of the eighteenth century, rival systems of nomenclature arose. By 1850, however, the system now in use, whereby a single name is used for each mineral, had been rather generally adopted.

To the question "How are minerals named?" the most straightforward answer is, "In any way the original describer wishes, with no requirement or system other than that whereby the name generally, but not always, ends in *ite.*" Several names are based on Greek or Latin words that provide some information about the mineral— for example, its color (albite, from the Latin *albus,* meaning white) or its high density (barite, from the Greek *barys,* meaning heavy). Other names relate to the mineral's composition—for example, calcite ($CaCO_3$) and zincite (ZnO).

These were admirable methods for naming minerals, because they gave some indication as to the nature of the mineral to which they were applied. Such names soon were used up, however, and most more recently named minerals have been named after localities (as aragonite for Aragon, Spain) and people (as joesmithite for the mineralogist Joseph V. Smith). [See R. S. Mitchell's interesting book about mineral names; it is cited at the end of this chapter.]

The accepted standard procedure for naming a newly discovered mineral has been set by the International Mineralogical Association. [See the 1970 article by Michael Fleischer in the references at the end of this chapter.] Currently accepted names for mineral species and varieties are given in Dr. Fleischer's *Glossary*, which has been updated periodically since its first publication in 1971.

The purpose of classifying minerals is to bring similar minerals together and to separate them from dissimilar minerals, thus giving them order. Today a chemical classification—generally attributed to the Swedish chemist Berzelius—is used throughout the world. The major classes in this classification are as follows:

Class	**Example** (Formula)
Native elements	Gold (Au)
Sulfides	Galena (PbS)
(+ selenides and tellurides)	
Sulfosalts	Tetrahedrite [(Cu,Fe)$_{12}$Sb$_4$S$_{13}$]
Oxides	Hematite (Fe$_2$O$_3$)
Hydroxides	Gibbsite [Al(OH)$_3$]
Halides	Fluorite (CaF$_2$)
Carbonates	Calcite (CaCO$_3$)
Nitrates	Nitratite (NaNO$_3$)
Iodates	No common one
Borates	Borax (Na$_2$B$_{407}$·10H$_2$O)
Sulfates	Barite (BaSO$_4$)
(selenates, *etc.*,)	
Chromates	Crocoite (PbCrO$_4$)
Phosphates	Apatite [*e.g.*, Ca$_5$(PO$_4$)$_3$F]
(arsenates, *etc.*..)	
Antimonates (*etc.*)	No common one
Vanadates	Carnotite [K$_2$(UO$_2$)$_2$(VO$_4$)$_2$·3H$_2$O]
(molybdates, *etc.*)	
Organic compounds	No common one
Silicates:	
Mesosilicates	Kyanite (Al$_2$SiO$_5$)
Sorosilicates	Epidote [Ca$_2$(Al,Fe)$_3$Si$_3$O$_{12}$(OH)]
Cyclosilicates	Tourmaline [NaFe$_3$Al$_6$(BO$_3$)$_3$Si$_6$O$_{18}$(OH)$_4$]
Inosilicates	Diopside (CaMgSi$_2$O$_6$)
Phyllosilicates	Muscovite [KAl$_2$(Si$_3$Al)O$_{10}$(OH)$_2$]
Tectosilicates	Orthoclase (KAlSi$_3$O$_8$)

The classes can be subdivided into groups; the native elements, for example, can be subdivided into the metals, the semimetals, and the nonmetals. The groups can be subdivided into families and/or species; and some species have one or more named varieties. In practice, the classification system must be treated with flexibility or certain essentially irresolvable taxonomic problems develop. Fortunately the International Mineralogical Association has a subcommission that deals with classification questions that arise, and perhaps even more fortunately, most such questions are of little, if any, interest to any but a few professional mineralogists.

REVIEW QUESTION

If you had a question about the proper naming of a mineral (for example, "Is sphene or titanite the accepted name for $CaTiSiO_5$?"), to what reference would you turn?

THE FORMATION OF MINERALS

Mineral formation is controlled by chemical (including biochemical) and physical conditions. In the main, the controlling factors are temperature, pressure, and the identity and proportions of the elements present. Several of the more common environments within which diverse minerals are formed are described in Chapter 3, "Rocks."

Most natural crystals, which are the most sought-after mineral specimens, grow where (1) the constituent atoms (or ions) are free to come together in the proper proportions; (2) the conditions are such that growth may occur at an appropriately slow and steady rate; and (3) the external surface of the growing crystal is not physically constrained.

As a consequence, most well-developed crystals are found lining the walls of open spaces within rocks—for example, in open fractures, solution cavities, and vesicles. Some such crystals have been deposited from hot aqueous solutions, often called *hydrothermal solutions*; others have been deposited as a result of the condensation of gases.

Crystals range in size from those that may be seen only with the aid of a microscope, up to, for example, a spodumene crystal (found in a pegmatite near Keystone, South Dakota) that was about 15 meters (nearly 50 feet) long. The microscopic crystals are often collected, appropriately mounted (see Figure 2–4), and assembled in "micromount" collections. The perfection of the crystal shape of microcrystals commonly surpasses that of larger crystals. Crystal surfaces, however, whatever the crystal size, range from mirror-smooth to rough and pitted or striated; and, both extremes may even be seen on some individual crystals (Figures 2–5A and 2–5B).

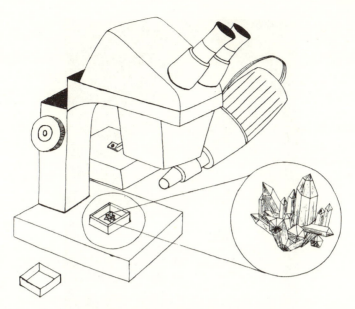

Figure 2–4   The collection, mounting, and study of micromounts is popular among many of today's collectors. As shown, these specimens are so small that one usually has to look at them with a microscope.

A

Figure 2–5.   The quality of crystal faces may differ greatly even on individual crystals. (A) Calcite from Belmont Quarry near Staunton, Virginia largest dimension ~ 30 centimeters (~ 12 inches). (Photograph courtesy of Smithsonian Institution.)
(*Continued next page*)

B

(B) Quartz crystal from Floyd County, Virginia.

REVIEW QUESTIONS

Would you be more likely to find a fine crystal in a solid rock or lining a cavit
within a rock? _____
Why? _____
_____
_____

Are relatively large or relatively small crystals more likely to exhibit more smooth
mirror-like faces? _____
_____
_____

THE OCCURRENCE OF MINERALS

Cavities that are only partially filled with minerals have supplied many of the finest mineral crystals now gracing the cabinets of both private collections and museums. Most of the cavities, which may become sites of crystal deposition, have names that are widely used by geologists and others who are interested in rocks, minerals, and mineral deposits. Here are a few of the more common ones:

Vesicles are cavities, typically spheroidal, formed by the entrapment of gas bubbles within the magma (molten rock material) that consolidated to form the rock (Figure 2–6). Zeolites, prehnite, and agate are especially common minerals in vesicles in basalt.

Figure 2–6. Vesicles in a basalt. See also Figure 3–13. (Photograph courtesy of Smithsonian Institution.)

Vugs are cavities, typically crystal-lined, in sedimentary rocks, especially in calcareous or dolomitic rocks. The term *druse* is frequently applied to the surfaces covered by small crystals projecting into the central void. The term *geode* is used if the filling is easily separable as a nodule from the containing cavity (Figure 2–7).

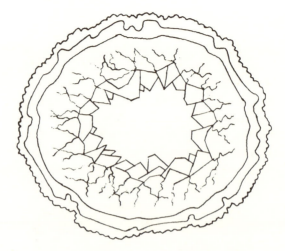

Figure 2–7. A typical geode, separable from the surrounding rock and crystal-lined.

Solution cavities are formed when fluids flowing through rocks dissolve the rocks. Caves and joints and bedding planes that have been widened by solution are examples. Later, solutions may deposit minerals in these cavities (Figure 2–8). A number of ore deposits with magnificent crystals of, for example, sphalerite and galena, as well as the better-known cave deposits, such as stalactites, have been found in such cavities.

A

Figure 2–8. Caves and enlarged joints are frequently the loci for the deposition of crystals. (A) Calcite group from a cave near Buchanan, Virginia. (continued on next page)

B

(B) Block-shaped barite crystals and a saddle-shaped group of dolomite crystals from a joint plane deposit near Glasgow, Virginia. (Photographs courtesy of G. K. McCauley and T. M. Gathright, Jr.)

Shear zones may constitute open fissures or breccias that contain much open space. The open spaces subsequently may be filled—either partially or wholly—to form veins. Those that are only partially filled commonly provide excellent mineral specimens.

Miarolitic cavities, especially common in granitic rocks, appear to represent loci where gas bubbles were trapped within the cooling magma. This gas enhanced the free growth of crystals inward from the consolidating magma. Especially fine crystals of rock-forming minerals such as feldspar and quartz have been found in these cavities.

Pegmatites, unlike the other features just described, do not, as a rule, have central cavities. The word pegmatite is used widely to refer to any abnormally coarse-grained rock with an overall interlocking (igneous-like) texture. Although most pegmatites are of granitic composition, other compositions ranging from periodotitic to syenitic (see Table 3.1) also occur. Most grains of pegmatites exceed a centimeter in their greatest dimension; some individual grains have been found to measure a few meters (several feet) across.

Individual pegmatite masses may consist of a rather homogeneous mixture of microcline perthite and quartz with or without smaller amounts of, for example, biotite mica. These are known as simple pegmatites. Other pegmatites are zoned and contain, in addition to the perthite and quartz, large amounts of albite (var. cleavelandite) and well-formed crystals of such minerals as apatite, beryl, topaz, and colored tourmaline. These are known as complex pegmatites.

Simple pegmatites generally occur as tabular masses called *dikes* and contain few, if any, of the crystals desired by collectors. On the other hand, most complex pegmatites are irregular, lens-shaped masses and many provide wonderful opportunities for collecting. In many cases, extremely well-developed crystals may be collected because they are in "pockets" and are easy to break loose from their surrounding minerals. In general, however, you should not try to force them; you might break an irreplaceable specimen. Also, the crystal not only may be more attractive, but also of more value, if left attached to its associated minerals.

Other places where minerals occur will become evident to you when you read Chapter 3. In fact, you may want to scan that chapter before completing this one.

Because temperature, pressure, and chemical conditions control mineral formation, minerals that form under the same general conditions may be expected to occur together, whereas minerals that require different conditions for formation are unlikely associates. Thus, a knowledge of mineral associations is a valuable tool for mineral collectors and prospectors.

REVIEW QUESTION

If you know that mineral X is formed only at high temperatures and that mineral Y is formed only at low temperatures, how might you account for the presence of crystals of Y on top of crystals of X in a vein? _____

_____

_____

THE IDENTIFICATION OF MINERALS

Each of the nearly 3000 known minerals may be identified by microscopic examination, X-ray diffraction, chemical analysis, differential thermal analysis, or some combination of these methods. Most of the relatively common minerals that you are likely to see, however, also may be identified in hand-specimen on the basis of their appearances or by subjecting them to a few simple tests. The general procedure is to match the properties of the "unknown" with those listed in determinative tables such as those given in Appendix II. The following properties are usually considered.

Luster is the appearance of a fresh surface of a mineral in reflected light. Most minerals can be readily characterized on the basis of their luster as either metallic or nonmetallic; a few are better described as submetallic. In addition, many minerals with nonmetallic lusters may be distinguished from one another by descriptive terms such as *adamantine* (brilliant, like a diamond), *vitreous* (glasslike), *pearly*, *silky*, and *dull*.

REVIEW QUESTION

List one or more materials (not necessarily minerals) with each of the following lusters.

Metallic: _____

_____

Nonmetallic: vitreous_____

silky _____

pearly _____

dull _____

earthy _____

Color, or the lack of color can help in identifying some minerals. Most minerals exhibit either *inherent color* or *exotic color*. Inherent color is color that depends upon the overall composition of the mineral—for example, the golden color of native gold. Exotic color depends upon the presence, commonly in only trace amounts, of some foreign material distributed throughout the mineral—for example, a pigment. Several of the colors seen in quartz, which is inherently colorless, are examples of exotic coloration. A few minerals exhibit other or additional color effects, such as a "play of colors."

Streak, which may be defined as the color of the powder of a mineral, is frequently cited as characteristic for some minerals, such as hematite and limonite. It may be quite different from the readily apparent color of the mineral in mass. In essence, observation of streak eliminates spurious effects attributable to differences in, for example, grain size. The streak may be attained by using a streak plate, by grinding a piece of the mineral in a mortar, or even by hammering and thus "bruising" some surface of the specimen.

In any case, until you have a lot of experience, it is neither necessary nor best to rely upon color or streak in identification. Note, for example, that several mineralogists and mineral collectors, who are highly skilled in mineral identification are color-blind.

REVIEW QUESTIONS

What are the two main types of color related to minerals? 1. _____
2. _____.

In the determinative tables of Appendix II, each mineral is given under all of its common colors. Examine the tables and list a few minerals of each type referred to in the preceding question.

1. _____   2. _____

_____   _____

_____   _____

_____   _____

Do a mineral's color and streak always match? _____
If not, which is more likely to be distinctive and thus of more use in identification?

Is a color-blind person necessarily incapable of engaging in mineral identification? Why or why not? _____

_____

_____

Hardness is the resistance of a mineral to abrasion or scratching. Although there are special devices for measuring hardness quantitatively, the property usually is measured only in relative terms—that is, whether the mineral is harder or softer than some other mineral or material. The hardness of a mineral is generally designated by a number on a scale first suggested by Friedrich Mohs in the 1820s. Mohs selected ten relatively common minerals to represent various degrees of hardness. These ten minerals, arranged in order of increasing hardness, are:

1. Talc
2. Gypsum
3. Calcite
4. Fluorite
5. Apatite
6. Orthoclase (feldspar)
7. Quartz
8. Topaz
9. Corundum
10. Diamond

To determine the Mohs hardness of a mineral, one tries to scratch fresh surfaces of materials of known hardnesses with a sharp corner of the unknown mineral, or vice versa. Each mineral, of course, will scratch materials softer than itself. For example, a mineral with a hardness (H) of 2½ will scratch gypsum or talc, but will be scratched by calcite and other materials with hardnesses of 2½ or greater.

Although "hardness pencils" (small rods with materials of known hardnesses attached to their ends) may be used to determine a mineral's hardness, most geologists estimate the approximate hardness by using materials readily at hand, and, with a little experience, one can estimate degrees of differences. The most frequently used materials are the fingernail ($H \cong 2\frac{1}{2}$); a copper coin ($H \cong 3\frac{1}{2}$); a jackknife, hammer, or piece of common window glass ($H = 5–5\frac{1}{2}$); and a piece of quartz, such as a sand grain ($H = 7$).

When determining hardness, three precautions must be kept in mind: (1) Do not mistake the powder of a softer mineral left on the surface of a harder material for a scratch on the harder material. (2) Do not confuse either the tearing apart of the grains of an aggregate or the breaking off of small cleavage fragments from a mineral as scratches. (3) Remember that a few minerals (for example, kyanite) have notably different hardnesses in different directions.

REVIEW QUESTIONS

Using the tables in Appendix II, list:

Two minerals that can be scratched with your fingernail. _____

_____

Four minerals that will scratch your fingernail but be scratched by a knife blade.

_____ _____ _____ _____

_____

Three minerals that will scratch glass but be scratched by quartz. _____

____ _____ _____

Why are diamond, corundum, and quartz good abrasives? _____

_____

Durability is desired for gemstones used in rings. Does Mohs' scale give any indication that hardness and durability are at least partially dependent upon each other? Why or why not? _____

Specific gravity is the ratio of the weight of a substance to the weight of an equal volume of water (strictly speaking, of water at 4°C and under 1 atmosphere pressure). The term *density*, used as a synonym for *specific gravity* by some people, is of equal numerical value when stated in terms of grams per cubic centimeter.

Although several devices have been made especially for determining the specific gravity of mineral specimens, it can be determined rather simply. A commonly used procedure (see Figure 2–9) consists simply of weighing the specimen in air; weighing it again, submerged in water; and substituting those weights in the appropriate places in the following equation.

$$\text{Specific gravity of } X = \frac{\text{Weight of } X \text{ in air}}{\text{Weight of } X \text{ in air} - \text{Weight of } X \text{ in water}}$$

The basis of this equation is the fact that the loss of weight in water is equal to the weight of an equal volume of water (Archimedes' Principle).

Specific gravity is especially useful when one wishes to identify a specimen without submitting it to a possibly destructive test, such as checking its hardness. On the other hand, it is inconvenient, especially in the field, because of the equipment required. Two methods have been used to help overcome this difficulty: (1) Some persons are able, with practice, to estimate specific gravity by hefting specimens. (2) Fluids of known specific gravities may be carried and used to check specimens. The specimens will sink in fluid if their specific gravity is greater than that of the fluid, whereas they will float if their specific gravity is less than that of the fluid, or remain at any position where they are placed if their specific gravity is the same as that of the liquid.

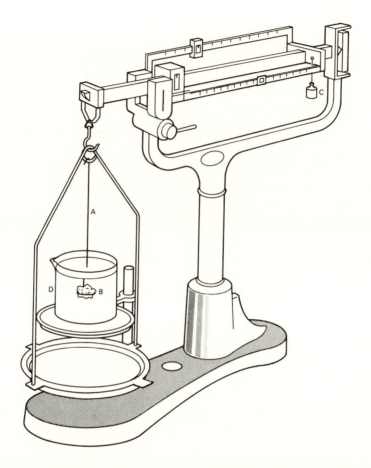

Figure 2–9. Using a typical laboratory balance to determine the specific gravity of a mineral sample. Three steps are required: (1) The balance with support wire A but without sample B is balanced at 0 grams with the counterweight C. (2) Sample B is placed in the support wire and weighed in air. (3) The beaker of water, D, is raised so the water completely covers the sample, and the sample is then reweighed.

REVIEW QUESTIONS

Why does the mineral ice float on its liquid counterpart, water? _____

Take a thumb-sized piece of some common mineral such as quartz or calcite and determine its specific gravity as shown in Figure 2–9. Compare the value you have obtained with that given for the mineral in Appendix II. _____

Crystal is the name given to a solid whose surfaces display the previously mentioned regular, periodic, three-dimensional arrangement of the solid's constituent atoms or ions. The study of crystal shapes is valuable in hand-specimen mineral identification because many minerals have shapes that are both characteristic and relatively common. All known crystal shapes belong to one of the crystal sytems depicted and described in Figure 2–10. The crystal system is given for each mineral included in the determinative tables in Appendix II.

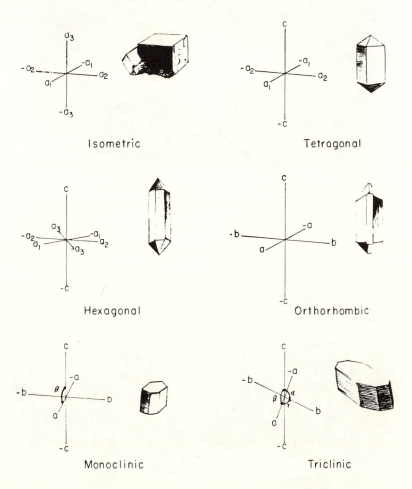

Figure 2–10. The six crystal systems. *Isometric system:* $a_1 = a_2 = a_3$, and all interaxial angles = 90°. *Tetragonal system:* $a_1 = a_2 \neq c$, and all interaxial angles = 90°. *Hexagonal system:* $a_1 = a_2 = a_3 \neq c$, angles between a-axes are 120°, and angle between plane of a-axes and c = 90°. *Orthorhombic system:* $a \neq b \neq c \neq a$, and all interaxial angles = 90°. *Monoclinic system:* $a \neq b \neq c \neq a$, and angle between a and c (β) is not a right angle, whereas angle between b and the a–c plane is a right angle. *Triclinic system:* $a \neq b \neq c \neq a$, and no interaxial angles are right angles.

In addition, some minerals commonly consist of several individuals with distinctive grouping, each with its own name. Some mineralogists refer to this descriptive term as the mineral's *habit*. A few of these terms that you are likely to encounter are the following.

Acicular refers to needlelike groups (Figure 2–11).

Figure 2–11.   Acicular calcite. (Photograph courtesy of Smithsonian Institution.)

Dendritic is used to describe plantlike forms (Figure 2–12).

Figure 2–12.   Dendrites. Tree-shaped deposits of manganese oxides are relatively common on some rock surfaces. (Photograph courtesy of Smithsonian Institution.)

Colloform (or sometimes *botryoidal*) describes groups characterized by external surfaces that are rounded or consist of roughly hemispherical prominences (Figure 2–13A). Where broken, colloform masses commonly exhibit either radiating or concentric structure or both (Figure 2–13B).

A

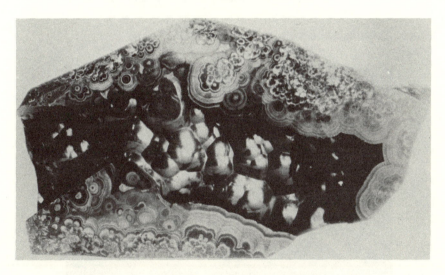

B

Figure 2–13. Colloform habit. (A) Colloform psilomelane from Crimora, Virginia length of specimen ~ 28 centimeters (~; 11 inches). (B) Colloform malachite from Zimbabwe exhibiting internal radiating and concentric patterns. (Photographs courtesy of Smithsonian Institution.)

Oolitic is applied to masses that resemble fish roe (Figure 2–14).

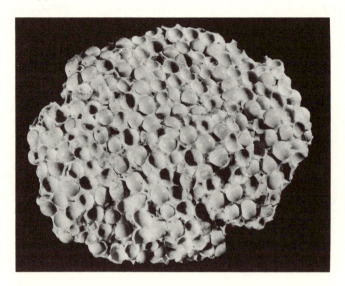

Figure 2–14.   Oolitic calcite. (Photograph courtesy of Smithsonian Institution.)

Stalactitic suggests iciclelike shapes (Figure 2–15).

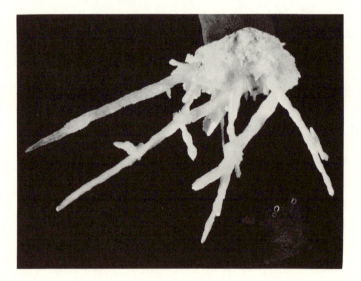

Figure 2–15.   Stalactitic calcite from Luray, Virginia.

Many other descriptive terms—such as *compact earthy, fibrous* (Figure 2–16), *platy,* and *bladed*—are used and their meanings are self-evident.

Figure 2–16. Fibrous serpentine asbestos.

REVIEW QUESTIONS

If you found a crystal in the shape of a child's building block (that is, with all square sides), to which crystal system would it most likely belong? _____
If you found a crystal that had a three-sided pyramid on top, to which crystal system would it belong? _____
If you found an eight-sided crystal shaped like two four-sided pyramids placed base to base, to which *three* crystal systems might it belong? _____

Name and briefly characterize three common mineral habits.

_____ : _____.
_____ : _____.
_____ : _____.

Cleavage refers to the tendency of certain minerals and other crystalline substances to break or split along certain plane surfaces whose positions are controlled by the regular internal arrangement of constituent unit cells ("building blocks"). Some minerals have no discernible cleavage. Some have a single direction, called a plane, of cleavage (Figure 2–17A); others have two or more planes of cleavage (Figure 2–17B). Parallel plane surfaces of cleavage represent the same cleavage direction. The different cleavage surfaces of specific minerals have characteristic angles between them—for example, halite (common table salt) has three directions of cleavage at right angles to each other.

A

B

Figure 2–17. Cleavage. (A) Mica, with one direction of cleavage, cleaves into sheets. (B) Calcite has three directions of cleavage *not* at right angles to each other. (Photographs courtesy of Ward's Natural Science Establishment.)

Fracture is the term usually applied to surfaces along which a mineral breaks when it does not cleave. Several minerals exhibit characteristic fractures, which are generally described by such terms as *conchoidal* (said of fractures that resemble the smooth curves of broken glass, Figure 2–18), *splintery, rough,* and *irregular.*

Figure 2–18. Conchoidal fracture of a piece of quartz. The curved fracture surfaces resemble those of broken glass.

REVIEW QUESTIONS

Look at several grains of table salt through a magnifying glass. What is the most common shape of the grains? _____. What is the angle between adjacent faces? _____. Pick out one of the larger grains; tap it with, for example, the handle of a knife. Does the grain break along one or more cleavage planes? _____. Was it (or were they) parallel to the plane (or planes) of the original grains? _____.
Break a piece of glass (carefully!). Look at the broken surface. Is it a cleavage or a fracture? _____. If a fracture, how would you name (characterize) it? _____.

Other properties that may aid in the identification of certain minerals include diaphaneity, magnetism, relative solubility (including taste), feel, odor, luminescence (for example, fluorescence and phosphorescence), radioactivity, and tenacity. A few examples are noteworthy.

> *Diaphaneity*:   Minerals may be transparent (capable of being seen through), translucent (capable of letting light pass through), or opaque (incapable of permitting light to pass through).

Magnetism: The mineral magnetite and most pyrrhotite are attracted to a magnet (Figure 2–19).

Figure 2–19.   Some minerals act like magnets. This is the variety of magnetite generally called lodestone.

Solubility: Halite and several other minerals are soluble in water and certain other liquids; for example, calcite dissolves with brisk effervescence (bubbling) in dilute hydrochloric acid (HCl).

Feel: Talc, graphite, and serpentine are unctuous, that is, they feel greasy or soapy.

Odor: Arsenopyrite and a few other minerals emit typical odors when broken, scratched, heated, or acted upon by certain solutions.

Luminescence: Some varieties of calcite, fluorite, and several other minerals give off characteristic colors under ultraviolet ("black") light, and thus are said to fluoresce.

Radioactivity: Uraninite and a few other minerals contain radioactive elements that emit charged particles that cause a Geiger counter to click or flash.

Tenacity: Because of differences in tenacity, some minerals are malleable (for example, gold) whereas others are brittle (for example, garnet); and, cleavage flakes of some minerals are elastic (for example, those of the micas), whereas those of others are not (for example, those of the chlorites).

And these are just a few examples of the other characteristics of minerals that may be used in identification.

REVIEW QUESTION

Considering, but not necessarily using, all of the properties described for minerals in this section, outline a procedure that might lead to the identification of many mineral specimens. _____

_____

_____

_____

_____

_____

_____

_____

_____

_____

_____

With your answer in mind, preview the procedure suggested in the introductory remarks at the beginning of Appendix II. The fact that your suggested procedure differs from the one given there does not necessarily mean that yours is not as workable—especially for you—as the one we have suggested!

MINERALOIDS, GLASSES, AND MACERALS

Three kinds of natural substances that are not minerals but commonly occur with minerals and are important and/or interesting geological materials are mineraloids, glasses, and macerals.

Mineraloids are typically inorganic and have many attributes of minerals, except that they are amorphous. The most commonly cited example is opal. Though not strictly amorphous, opal is exemplary in that it is thought to be formed from a gel.

Glass, which is amorphous, is formed when a molten rock material—for example, a magma—is quenched. The rapid cooling produces a supercooled liquid that lacks the ordered internal structure required, by definition, for a mineral. The apparent rigidity of glass is merely an expression of extreme viscosity (*viscosity* is the opposite of fluidity). Several types of rocks formed by the relatively rapid cooling of magma consist partly, or even predominantly, of natural glass. Obsidian is the most common natural glass (Figure 2–20). Natural glass is also produced by other processes, as, for example, when lightning strikes moist sand or when large meteorites impact certain rock materials.

0   2 cm

Figure 2–20. Obsidian (volcanic glass) from Yellowstone National Park, Montana, exhibiting typical conchoidal fracture. (Photograph courtesy of B. J. Skinner.)

Macerals are the microscopically discernible organic unit materials—such as woody tissue, spores, and fossil charcoal—that constitute coal and portions of several other rocks. It is generally considered that macerals are to coal as minerals are to rock. Whereas most mineral names end in "ite," most maceral designations end in "inite." Macerals are derived from the maceration of pieces and products of vegetation. Unlike minerals, whose nomenclature is rather closely controlled, the names of some of the macerals are rather confused because coal petrographers, organic geochemists, and micropaleontologists all tend to apply at least some of the terms differently. General definitions of terms are given in Berry *et al.* (see the references at the end of this chapter).

REVIEW QUESTIONS

Crystalline substances yield X-ray patterns because of the ordered internal arrangements of their constituent atoms. Would you expect mineraloids and glasses also to yield X-ray patterns? _____. Why or why not? _____

_____

_____

On the basis of economic considerations, do you think aspiring geologists might better study mineraloids, natural glass, or macerals? _____. Why?

_____

_____

REFERENCES AND SUGGESTED ADDITIONAL READING

Berry, L. G., Mason, B., and Dietrich, R. V., *Mineralogy: Concepts, Descriptions, Determinations,* 2nd ed. San Francisco: W. H. Freeman, 1983.

Fleischer, M., Procedure of the International Mineralogical Association Commission on New Minerals and Mineral Names. *American Mineralogist,* vol. 55, pp. 1016–1017, 1970.

Fleischer, M., *Glossary of Mineral Species.* Tucson, Ariz.: Mineralogical Record, 1980.

Mitchell, R. S., *Mineral Names[–]What Do They Mean?* New York: Van Nostrand Reinhold, 1979.

Pough, F. H., *A Field Guide to Rocks and Minerals.* Boston: Houghton Mifflin, 1955.

Zim, H., and Shaffer, P., *Rocks and Minerals* (Golden Nature Guide). Racine, Wis.: Western Publications, 1957.

CHAPTER 3

Rocks

Rock-to-mineral relationships bring several analogies to mind: Rock is to mineral as forest is to a lonesome pine; as an exquisite tapestry is to a piece of silk thread; as a spring bouquet is to a daisy . . . Certain forest-to-tree relationships clarify a number of the diverse aspects of rock-to-mineral relationships rather well: Although many natural forests contain several species of trees and most rocks contain several different minerals, some forests are made up mainly of numerous trees of a single species, just as some rocks are composed largely of many grains of a single mineral. Also, just as the trees of a forest may be of different shapes and sizes, so may the mineral grains of a rock be of many shapes and sizes. And, although some forests are similar and may be classified together, each is truly unique, just as each rock is unique. (This relationship is evident in Figure 3.1.)

Dietrich, R. V., *Stones: Their Collection, Identification, and Uses*. San Francisco: W. H. Freeman, 1980.

In this chapter, you first learn the definition of *rock* and then examine the diverse aspects of that definition. Then you learn how rocks are named and classified, how they are formed, how they occur, how they may be interrelated, how they are described and identified, and what rocklike materials are really pseudorocks.

REVIEW QUESTIONS

List some ways in which rocks differ from minerals. _____
_____
_____
_____

Give some ways in which you have seen rocks used (for example, in the building industry). _____
_____
_____

Of the things you will learn in this chapter, which do you expect to find most interesting? _____
_____

(Check this answer after you finish the chapter to see if you have changed your mind.) _____

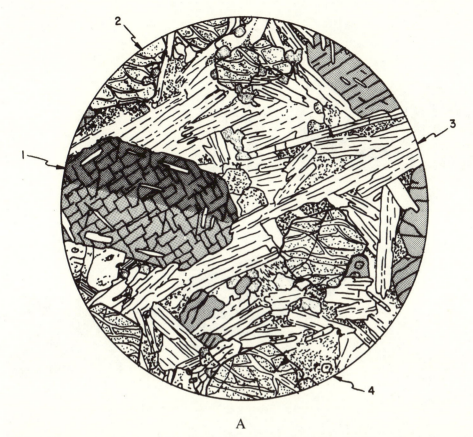

A

Figure 3–1. (A) Rock slice (thin section) with minerals represented as follows: (1) pyroxene; (2) olivine; (3) plagioclase feldspar; and (4) apatite with magnetite inclusions. As an example of how this rock might appear through a microscope between crossed polars: the pyroxene and olivine could be any color, dark gray, or any combination of such colors; the plagioclase would be black to white with alternate lamellae ranging from black and white to essentially equivalent gray hues; the apatite would be dark gray to nearly white; and the magnetite would be black (opaque). (Reproduced by permission from R. V. Dietrich, *Geology and Virginia,* © 1970 University Press of Virginia.) (*Continued on next page*)

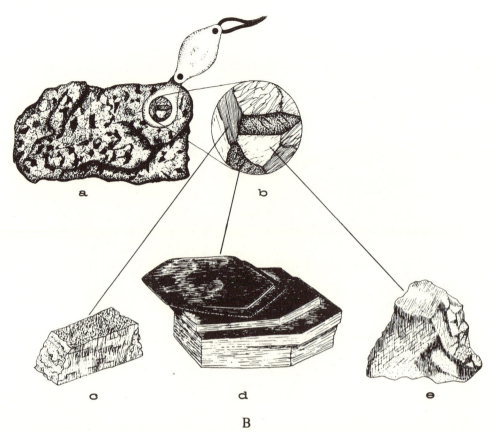

B

(B) (a) A rock—for example, granite. (b) The rock as seen with aid of a hand lens. (c, d, e) Specimens of the chief mineral constituents of the rock. (c) Feldspar with two cleavages at nearly right angles. (d) Mica with one cleavage (which yields elastic plates). (e) Quartz with no cleavage but conchoidal fracture.

DEFINITION OF *ROCK*

A rock may be defined as a natural "solid" composed of mineral grains, mineraloids, macerals, and/or glass.

There are three chief aspects to this definition. A rock must:

1. Be natural
2. Be "solid"
3. Consist of one or more of the following.
    a. mineral grains
    b. mineraloids
    c. macerals
    d. glass

Each of these requirements warrants further consideration.

Natural, as in the definition for mineral, simply means formed in nature. This requirement excludes manufactured materials, such as concrete and slag, that otherwise fit the definition.

"Solid," as applied to rock, must be enclosed in quotation marks, so long as obsidian and other natural glasses are considered rocks. In addition, the "solidity" of rocks is rather subjective when applied to, for example, materials that range from masses of loose fragments (such as sand) to their well-cemented equivalents (such as sandstone). A general rule is this: If you need a hammer to break it, it is a rock.

The reason for including all the diverse materials—mineral grains, mineraloids, macerals, and glass—is that any one of these natural substances may constitute a rock. The reason "and/or" is used in the definition is that essentially any combination of the constituents may be present in a given rock. For example, coals commonly consist of macerals and mineral grains, and many volcanic rocks consist of both mineral grains and glass. The reason for using "mineral grains" in the definition instead of merely "minerals" is that some rocks consist essentially of grains of only one mineral rather than a combination of two or more minerals. Even these so-called monomineral rocks, however, have appearances that distinguish them from specimens generally considered to be mineral specimens; stated simply, most rocks appear heterogeneous because of their composition and/or their textures (that is, the arrangement of their constituents), whereas most mineral specimens appear homogeneous. Some natural glasses, anthracite ("hard coal"), and extremely fine-grained rocks are exceptions to this generalization.

REVIEW QUESTIONS

Why shouldn't concrete, which is made up of sand grains that have been cemented together, be considered to be a rock? _____

Why should the word "solid" be enclosed in quotation marks when applied to obsidian and other natural glasses? _____

_____

_____

Statuary marble, such as that used in the Lincoln Memorial, is a rock that consists almost completely of the mineral calcite. Why is such marble a rock rather than a mineral? _____

_____

THE NAMING AND CLASSIFICATION OF ROCKS

Just as for mineral names, the origins of many rock names are also lost in antiquity. And, like minerals, rocks have been named after all sorts of things and people. A few examples may be given: The name limestone refers to its use in making lime (CaO). Phonolite comes from the Greek *phon,* meaning voice, because the rock rings when hit with a hammer. Phyllite is from the Greek *phyllon,* meaning leaf, in reference to the way the rock tends to break off in thin, leaflike pieces. Syenite comes from Syene (Asswan), Egypt. Charnockite gets its name from Job Charnock (or, more correctly, from his tombstone, which is made from that rock).

There is no widely accepted authority for rock names. Unfortunately, there also is no way to tell, without consulting a good reference book, such as the American Geological Institute's *Glossary of Geology* or Michael Fleischer's *Glossary of Mineral Species,* whether an unfamiliar name, such as yamaskite or gesundheit, refers to a rock, to a mineral, or to neither.

In most cases, the general classification of a rock is at least a subconscious step toward identifying and thus naming the rock. This is so because the overall classification of rocks is based on rock origin, which is generally reflected in the rock's composition, its textural appearance, or both. Consequently, the classification system will be given in this book after you find out about how rocks are formed and how they occur.

REVIEW QUESTION

Although you may not know many rock or mineral names, you very likely do know whether some of the names listed here are correctly applied to rocks or to minerals. Put "R" in front of those you think are rocks, "M" in front of those you think are minerals, and "X" in front of those that you think are neither rocks nor minerals.

1.	___	basalt	13.	___	marble
2.	___	calcite	14.	___	mica
3.	___	chromium	15.	___	obsidian
4.	___	coal	16.	___	opal
5.	___	concrete	17.	___	plasticine
6.	___	conglomerate	18.	___	polyurethane
7.	___	diamond	19.	___	pyrite ("fool's gold")
8.	___	feldspar	20.	___	quartz
9.	___	flint	21.	___	ruby
10.	___	gold	22.	___	sandstone
11.	___	granite	23.	___	sapphire
12.	___	limestone	24.	___	slate
			25.	___	steel

answers

(1, 4, 6, 9, 11, 12, 13, 15, 22, and 24 are "R." rocks; 2, 7, 8, 10, 14, 19, 20, 21, and 23 are "M." minerals; 3, 5, 16, 17, 18, and 25 are "X." neither.)

THE FORMATION OF ROCKS

The most commonly used classification of rocks is based on rock genesis,—that is, how rocks are formed. Thus, you may best understand how rocks are formed by considering the major classification categories. Most geologists use the tripartite classification of *igneous, sedimentary,* and *metamorphic* rocks. This classification, however, must be recognized as an oversimplification; several rocks fit much better along the boundary zones between two of the categories than within either one. Therefore, the classification is truly useful only if its shortcomings are kept in mind.

Igneous rocks are formed when molten rock material, called *magma,* is cooled sufficiently to be consolidated. (The term *magma* is used by some geologists for molten rock material below the earth's surface, and the term *lava* is used for magma that has been ejected out onto the surface.) The cooling may be rapid and result in the formation of glass, or it may be slow and result in the crystallization of mineral grains. As a general rule, the more slowly a magma is cooled and consolidated, the coarser its grain size.

Most natural glasses are of igneous origin; most crystalline igneous rocks have interlocking mineral grains (Figure 3–2). Differences in the compositions of magmas, along with changes that occur in the magmas as they move and are cooled, result in the great diversity of igneous rocks. As you will learn in the section dealing with the description and identification of rocks, the names of igneous rocks refer to certain compositions *and* grain sizes.

Figure 3–2.   An igneous rock, characterized by interlocking mineral grains, as you might view it through your hand lens.

Sedimentary rocks have two principal modes of formation: *lithification* (that is, turning to rock) of loose detrital fragments that have been transported and deposited on the surface, and *precipitation* of minerals from natural aqueous solutions on or near the earth's surface. Examples of sedimentary rocks formed by lithification are sandstones and conglomerates (cemented sands and gravels); examples of rocks formed by precipitation are rock salt and travertine (the name given to, for instance, rocks formed as stalactites and other cave deposits). Characteristics most commonly found in sedimentary rocks are layering (frequently termed *bedding* or *stratification,* Figure 3–3) and the presence of fossils (see Chapter 4). In addition, the fragmental nature of detrital rocks is clearly an indication of sedimentary origin (Figure 3–4). Most rocks formed by precipitation consist of interlocking grains, similar to those of coarse-grained igneous rocks (Figure 3–2); their mineralogical compositions, however, are generally indicative of their having been deposited as sedimentary rock from surface or near-surface aqueous solutions.

Figure 3–3.   Sedimentary layering in Upper Ordovician rocks on Walker Mountain, Virginia. (Photograph courtesy of T. M. Gathright, Jr.)

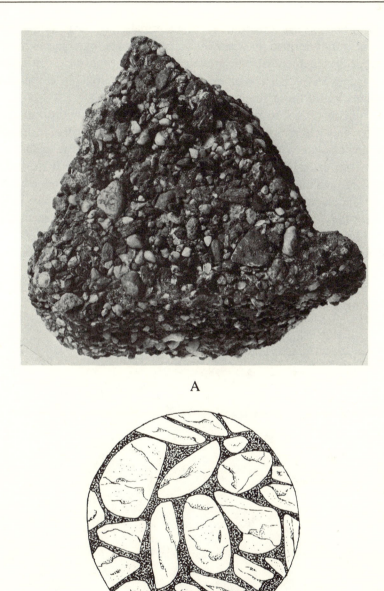

A

B

Figure 3–4. Conglomerates are cemented gravels—that is, they consist of detrital fragments that clearly indicate a sedimentary origin. (A) Photograph. (B) Sketch. (Photograph courtesy of Smithsonian Institution.)

Metamorphic rocks are formed by the transformation of preexisting rocks. The processes, generally termed *metamorphism,* are manifest by changes in the size, shape, and/or arrangement of constituents; changes in the rock's mineralogical makeup; changes in overall chemical, as well as mineralogical, composition; or some combination of these changes.

Metamorphism occurs in response to differential and/or hydrostatic pressure (Figure 3–5), elevated temperatures, and/or changes in chemistry. It may affect a relatively small volume of rock, such as that within a fault zone or that near the contact between a magmatic mass and the adjacent rocks, or it may affect an extremely large volume of rocks, such as that now constituting much of the Crystalline Appalachians of eastern North America. (*Fault zones* are loci where masses of rocks have been fractured and the originally adjacent parts moved with relation to each other.) Metamorphism within fault zones is often termed *dynamic* or *dislocation metamorphism;* that caused by the heat and/or the volatiles (fluids) given off by cooling magmas is frequently referred to as *contact* or *igneous metamorphism;* and that affecting large volumes of rock is generally termed *regional* or, in some cases, *dynamothermal metamorphism.*

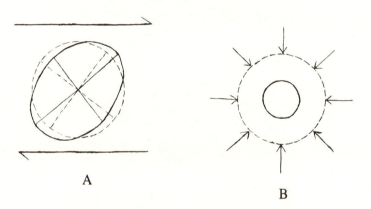

A

B

Figure 3–5.   Differential (A) versus hydrostatic (B) pressure.

It is generally assumed that metamorphic changes take place within essentially solid rock. Perhaps the most easily recognized characteristic of metamorphism is *foliation,* which is the macroscopically visible preferred orientation of rock constituents caused by their development under the influence of directed pressure. Many, but not all, metamorphic rocks exhibit such foliation (Figure 3–6).

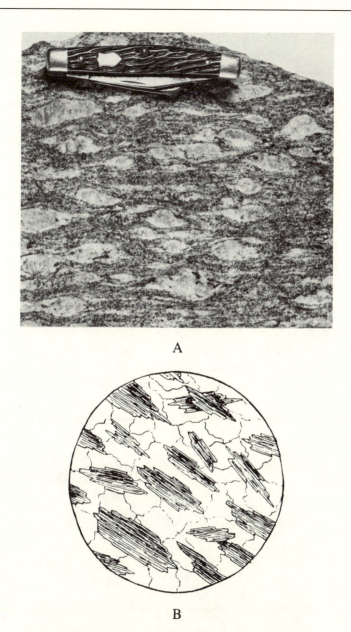

A

B

Figure 3–6.   Foliation. Many metamorphic rocks exhibit a preferred orientation of one or more of their mineral components. (A) An augen gneiss. (B) A biotite gneiss-schist.

Other rocks, those that do not fit well into the three main categories, include such things as meteorites (rock material from extraterrestrial sources), fulgurites (glass produced by lightning), and rocks produced by weathering. Among the most common other rocks are the following:

Pyroclastic rocks are formed by lithification of tephra deposits, which is the name given to loose materials such as ashes and blocks that are ejected during volcanic eruptions. From the standpoint of derivation, these rocks are igneous; from the standpoint of their deposition (and, in many cases, also their lithification), they are sedimentary.

Diagenetic rocks are sediments that have undergone marked chemical and/or physical changes that took place before and frequently contributed to their lithification. The changes, termed diagenesis, include all chemical and physical modifications and transformations of sediments other than those attributable to metamorphism and subaerial weathering. Diagenetic processes, which include solution, deposition, replacement, and recrystallization, are responsible for such rocks as recrystallized limestones, replacement dolostones and cherts and coals. These rocks are frequently considered sedimentary rocks; more correctly, they straddle the boundary zone between sedimentary rocks and metamorphic rocks.

Migmatites are mixed rocks that may be seen macroscopically to consist of some dark-colored metamorphic rock (for example, amphibolite or biotite gneiss) and a lighter-colored component, typically of granitic composition. In most migmatites, the darker-colored metamorphic component appears to have remained relatively immobile while the granitic component was relatively mobile (Figure 3–7). Be-

Figure 3–7. Migmatite. Polished block of the Morton Gneiss from the Minnesota River Valley.

cause of this appearance, these rocks have often been considered to fit in the boundary zone between igneous and metamorphic rocks. Strictly speaking, however, some migmatites may be wholly metamorphic.

Veins are mineral masses that have been deposited from hot, aqueous (hydrothermal) solutions in tabular to sheetlike fractures in rock (Figure 2–8). Veins range in size from microscopic to several meters (or yards) in thickness and up to several kilometers (a few miles) in length; they may be fairly uniform in thickness or differ greatly over their lengths. Individual veins may contain only one mineral or two or more minerals. Quartz veins and calcite veins, with or without other components, are especially common. The vein minerals may have sharp contacts with the surrounding rock or there may be an intervening transitional zone having characteristics indicating that the original parent rocks have been altered to or replaced by the rock now present. The fractures in which the vein minerals were deposited may represent several different kinds of features, such as joints, faults, solution-widened bedding planes, or cavities within breccias.

Weathering products that are solid may be termed rocks. The most common of these rocks are residual clays, laterites and bauxites, and gossans. The processes responsible for formation of these rocks include solution, oxidation, reduction, and hydration. During these processes, minerals of the parent rock material may be decomposed, new minerals may be formed, and mineral matter may be subtracted or added by through-flowing solutions. Macroscopically, the individual minerals of most of these rocks can only be identified under such "sack terms" as clay minerals, bauxite, and limonite.

REVIEW QUESTIONS

Why is the frequently used classification of rocks into igneous, sedimentary, and metamorphic rocks an oversimplification? _____
_____

What does the grain size of an igneous rock tell you about the speed with which its parent magma cooled? _____
_____

What are the two most common ways in which sedimentary rocks are formed?
_____

What are two features common in sedimentary rocks but not in igneous or metamorphic rocks? _____
_____

Would you expect either or both of these to occur in pyroclastic rocks? _____
_____

What are the three main controls of metamorphism?_____
_____   _____
Name and briefly describe three other kinds of rocks. _____
_____
_____
_____
_____
_____

THE OCCURRENCE OF ROCKS

Rocks make up the crust of the earth. Bedrock is the name applied to those rocks that may be seen at the surface in both natural and man-made exposures, such as cliffs and rock quarries, and also to those rocks that are directly beneath unconsolidated overburden, such as soil and loose sediment. In places where bedrock is exposed, observing and collecting rocks is simple; in areas where there is, for example, deep chemical weathering or deep burial by sediments, it may be nearly impossible to find any rocks other than those brought in for use in construction or for memorial monuments. Fortunately, in some areas like this, stones made up of diverse rock types are present in the overburden and thus readily available for observation and collecting. Some gravel pits in glacial deposits and some beaches covered with cobbles are indeed rock collectors' "heavens."

Throughout much of the world, the best way to find out where different kinds of rocks occur as either covered or exposed bedrock is to examine geologic maps of the area or areas in which you are interested. Geologic maps of diverse scales and types are available in many public and university libraries and elsewhere (see Appendix I).

While looking at geologic maps, you will see that many large regions are underlain by rocks of similar origins; for example, most of the folded Appalachians and midcontinent area of the coterminous United States have sedimentary rocks as bedrock, whereas much of the Adirondack region of New York State and the central part of Canada extending out from Hudson Bay are underlain by igneous and metamorphic rocks. Consequently, unless you live in or near a border area between two or more geological provinces, you may need to travel to see and collect different kinds of rocks. In any case, you are likely to find your time much better spent if you include the checking of maps (or checking with people who know) in your pretrip plans.

REVIEW QUESTIONS

If you live in an area covered with deposits from one of the great glacial ice sheets during the most recent Ice Age and you wish to make a collection of many different kinds of rock, how would you probably be able to do this with the least expenditure of time, energy, and money? Why? _____

_____

_____

_____

If you are looking at a map showing bedrock geology, what does the map actually tell you? _____

_____

Would such a map definitely indicate the rock, or rocks, you might be able to collect within the area it covers? Why or why not? _____

_____

_____

_____

INTERRELATIONSHIPS AMONG ROCKS: THE CRUSTAL "ROCK CYCLE"

The definition of metamorphic rocks clearly indicates their derivation from other rocks, including igneous rocks, sedimentary rocks, and preexisting metamorphic rocks. Although the definitions given for the other main kinds of rocks do not relate them to each other or to metamorphic rocks, they could. Examine Figure 3–8, and then consider the following group of questions. The answers can be found by analyzing the diagram, and in doing so, you will gain a useful, overall understanding of the so-called "Rock Cycle."

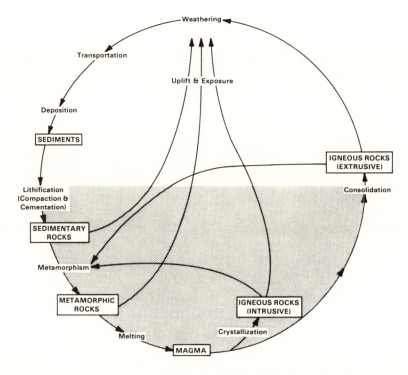

Figure 3–8.   The "Rock Cycle." Materials are shown in boxes; processes are not. Materials and processes shown in shaded, lower half of circle are formed or occur below the surface of the earth; others are formed or take place on or near the earth's surface. (Reproduced by permission from R. V. Dietrich, *Stones: Their Collection, Identification, and Uses.* © 1980 W. H. Freeman and Company, San Francisco.)

REVIEW QUESTIONS

The fragments derived by weathering, and transported, deposited, and then lithified to form sedimentary rocks may have previously been parts of what four kinds of rocks? _____

_____

Prior to lithification, loose fragments might undergo more than one period of transportation and deposition. Indicate this relation by adding another line, with an appropriate arrowhead, to the diagram.

Could chemically precipitated sedimentary rocks also be derived by chemical weathering of the four kinds of rocks given in your answer to the first question in this group? Why or why not? _____

In rare cases, nonmetamorphosed sedimentary rocks may be melted and the resultant magma might, in turn, be consolidated to form an igneous rock. As the diagram shows, a more likely thing is for metamorphosed rocks to be melted and thus form a magma, which may then be consolidated to form an igneous rock. Why is the first situation rarer than the second? _____

In what sequence of events might you expect a rock to undergo more than one metamorphism? _____

Explain how the "Uplift and Exposure" and "Weathering" indications on the diagram account for the fact that rocks formed deep below the earth's surface may constitute bedrock (both covered and exposed). _____

THE MACROSCOPIC DESCRIPTION AND IDENTIFICATION OF ROCKS

Igneous Rocks: Once an igneous rock has been identified, its name, together with a few descriptive adjectives, serves to describe it to the point that another person familiar with igneous rock terminology would know pretty much how the rock would appear. This is so because igenous rock names are based on the mineralogical composition and grain size of the rocks (see Figure 3–9 and Table 3.1). For example, the black and white (overall light gray) granite from Vermont that is widely used for tombstones (Figure 3–10) would be called a *light gray, medium-grained, biotite granite,* which means that it is overall light gray in color, that its main minerals have a greatest dimension ranging from about 2.0 to 5.0 millimeters, and that it contains noteworthy biotite (black mica) as well as the percentages of quartz, alkali feldspar, and plagioclase that lead us to classify it as a granite.

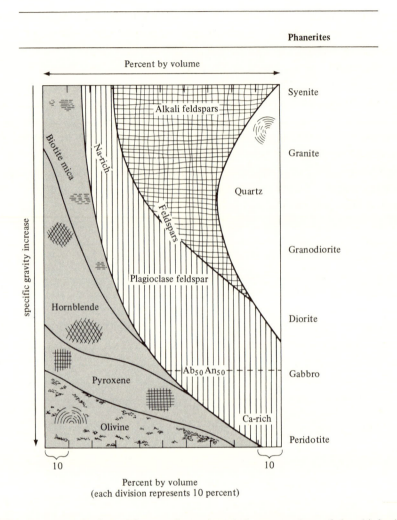

Percent by volume
(each division represents 10 percent)

Figure 3–9. Igneous rock chart. Diagram shows the relative proportions of the chief mineral components of common igneous rocks. Adjacent rocks listed in the column, to the right, grade into one another. The $Ab_{50}An_{50}$ compositional division in the plagioclase field cannot be determined macroscopically. (Modified after R. V. Dietrich, *Geology and Virginia.* © 1970 University Press of Virginia.)

Three sets of names are currently applied to the three diverse categories of relatively common igneous rocks. The set given on Figure 3–9 is used for *phanerites,* those igneous rocks in which the main minerals can be identified macroscopically; a second set is used for *aphanites,* igneous rocks in which the minerals are too small to be identified without a microscope; the third pair is used for *glasses.* The names of the three groups are correlated according to composition in Table 3.1.

Table 3–1 Common Igneous Rocks. The Chemical Compositions for Rocks on the
Same Horizontal Lines Are Essentially the Same. For Example,
Rhyolite is the Aphanitic Equivalent of Granite.

Phanerites	Aphanites		Glasses

Syenite
  (nepheline syenite)
	Felsite*	Trachyte
		  (phonolite)	

Granite		Rhyolite	Obsidian
Granodiorite		Dacite	
Diorite		Andesite	
Gabbro	Basalt		Tachylyte
Peridotite	None		

*If rock is completely aphanitic (that is, without phenocrysts), it is best to call
it a felsite if it is light-colored and a basalt if it is dark-colored.

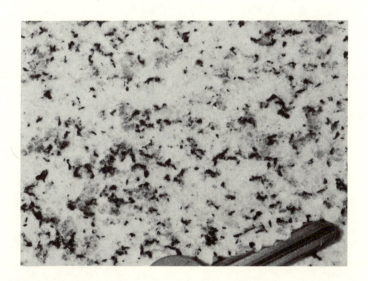

Figure 3–10.   Granite. Polished surface of a light gray, medium-grained, biotite granite.

You may wonder why two subsets of names are given for the aphanites,
especially since, by definition, their mineral components are too small to be
identified macroscopically. The reason is that in practice you may find each to be
useful. If the rock you are trying to identify is aphanitic throughout, it is best to call

it a felsite if it is light-colored and a basalt if it is dark-colored. [In this context, *dark-colored* means black, dark gray, or greenish hues of either of these, whereas *light-colored* pertains to all other colors, including such generally termed "dark" colors as dark reddish brown]. But, many aphanitic rocks can be named more specifically on the basis of macroscopic examination. This is so because they are *porphyritic*, that is, they contain phenocrysts, which is the name given to isolated grains that are large enough to be identified macroscopically (Figure 3–11). And, these relatively large grains are typically of the same identity and are commonly present in the same general proportions as the microscopic groundmass that surrounds them. Consequently, they indicate the overall composition of the rock and thus the specific name may be applied. For example, a rock with quartz, alkali feldspar, and plagioclase phenocrysts in the same proportions as those minerals in the already mentioned Vermont granite would be named *rhyolite* (or, more properly, *rhyolite porphyry*).

Figure 3–11. Porphyry. This small cobblestone is a trachyte porphyry from the Oslo district of Norway.

[It should be noted that the term porphyry is not restricted in its use to igneous rocks with aphanitic groundmasses. Rocks made up of large phenocrysts surrounded by a phaneritic groundmass are also porphyries. In any case, the widely accepted rule for naming a porphyry is to use the rock name that applies to the groundmass plus the designation porphyry. Thus, for example, a rhyolite porphyry and a granite porphyry might have essentially the same mineralogical composition but the rhyolite porphyry would have an aphanitic groundmass whereas the granite porphyry would have a phaneritic groundmass (Figure 3–12).]

Figure 3–12. Porphyry. A granite porphyry from the glacial drift in central Michigan.

Most igneous glasses you are likely to see would be called simply obsidian in the field. Chemical analysis would permit further identification as, for example, a trachytic obsidian. Tachylyte is the other at all common igneous glass. Two simple tests permit distinction between obsidians and tachylytes: obsidians are translucent in sliver-thin pieces, whereas tachylytes are opaque; tachylyte is readily soluble in HCl, whereas obsidian is not.

As is evident from this discussion, to identify and name an igneous rock on the basis of macroscopic examination, you need only determine the rock's grain size, to identify its chief mineral components, and to estimate the components' relative percentages. The common igenous rocks are included in Table 3–1 and in the identification tables in Appendix III. Several additional igneous rocks are described in *Rocks and Rock Minerals* by Dietrich and Skinner (see References at the end of this chapter).

A few additional features of igneous rocks are so common that you should know how to recognize them. And, you should also use the appropriate adjectives to indicate their presence in rocks. *Pegmatite,* and thus *pegmatitic,* have already been treated (page 33). *Vesicular* refers to igneous rocks, most commonly basalts, that contain spheroidal holes (vesicles) formed by the expansion of gas bubbles included within the parent magma during its consolidation (Figure 2–6). *Amygdaloidal* refers to vesicular rocks whose vesicles have been filled (Figure 3–13). The individual vesicles are *amygdules*.

Figure 3–13.   Amygdaloidal andesite. See also Figure 2–6.

REVIEW QUESTIONS

Name the following rocks on the basis of the descriptions given.

A phanerite consisting predominantly of alkali feldspar plus less than 10 percent quartz and about 10 percent hornblende._____

An aphanite porphyry of the same composition. _____

A dark gray aphanite containing many spheroidal holes. _____

A dark gray glass that can be seen to be translucent along thin edges. _____
_____

A greenish gray phanerite consisting of between 60 and 70 percent olivine.
_____

A medium-grained phanerite consisting of 50 percent alkali feldspar, 12 percent plagioclase, 30 percent quartz, and 8 percent biotite. _____

A fine-grained rock consisting of about 50 percent dark-colored plagioclase, 42 percent pyroxene, and 8 percent olivine, and having included spheroidal masses of diverse minerals, chiefly zeolites. _____

Sedimentary Rocks: Just as is true for igneous rocks, the name of a sedimentary rock, with or without one or more appropriate adjectives, serves to describe the rock fully to anyone familiar with sedimentary rock terminology. As you might expect, however, the bases for the names of sedimentary rocks are different from those for the names of igneous rocks.

For sedimentary rocks formed by the lithification of detrital fragments, the classification scheme is based on the grain size of the predominant fragments (see Table 3.2). As noted in the footnotes to that table, a few names (for example, shale and arkose) involve additional restrictions. Mixtures may be named as indicated in Figure 3–14.

Table 3–2   Detrital Sediments and Sedimentary Rocks.

Diameter of Fragments, Millimeters	Loose Aggregates	Consolidated Aggregates (Rocks)	Remarks
> 2	Gravel	Conglomerate	Rounded fragments
	Rubble	Breccia*	Angular fragments
1/16 - 2	Sand	Sandstone	Typically largely quartz†
1/256 -1/16	Silt	Siltstone	
< 1/256	Clay	Claystone	Includes shale‡ and mudstone

*Many breccias are not of sedimentary origin.
†Sand-size sedimentary rocks made up of 25 or more percent feldspar are called *arkose*; those with 25 or more percent dark minerals or rock fragments are called *greywacke*.
‡If fissile, that is, if it splits easily along bedding planes.

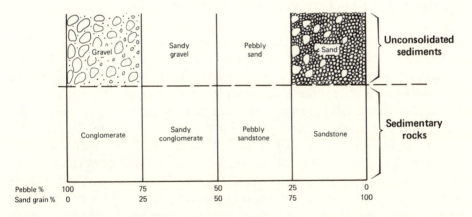

Figure 3–14. Names that may be used for mixed detrital sediments and sedimentary rocks. Mixtures other than those shown on the diagram would be named in a similar fashion. For example, a sedimentary rock consisting of 25 to 50 percent clay and 50 to 75 percent silt would be a clayey siltstone. (Reproduced by permission from R. V. Dietrich and B. J. Skinner, *Rocks and Rock Minerals.* © 1979 John Wiley & Sons, Inc., New York.)

For nondetrital sedimentary rocks formed by precipitation, either chemically or biochemically controlled, the basis for each name is mineral composition (see Table 3–3). Some of these rocks are given different names by different geologists—for example, *rock gypsum* rather than *gyprock* and *rock anhydrite* rather than *anhydrock*.

Table 3–3   Common Nondetrital Sedimentary Rocks.

Rock Name	Chief Mineral Constituent
Limestone*	Calcite
Dolostone†	Dolomite
Chert†	Cryptocrystalline (submicroscopic+) quartz
Chalk	Calcite and/or aragonite‡
Travertine	Calcite and/or aragonite§
Gyprock	Gypsum
Anhydrock	Anhydrite
Rock salt	Halite

*Some limestones consist of fragments and are better termed calcirudite (predominantly gravel-sized fragments), calcarenite (sand-size), or calcilutite (clay- or silt-size), as appropriate, see page 71.
†Many dolostones and cherts are formed by diagenetic replacement of calcium carbonate mud.
‡Largely powderlike and consisting of diverse mixtures of microorganisms.
§Solution-deposited in caves and around springs and seeps.

Although many geologists consider limestones to belong to the precipitate group, others think that many limestones warrant the establishment of a third category of sedimentary rocks, a group that may be termed *intrabasinal clastics*. These rocks are made up of calcium carbonate fragments that have been formed by chemical or biochemical precipitation, have been moved within the basin where they were precipitated, and subsequently have been lithified as a result of cementation. These rocks may be named on the basis of the identity of their constituent fragments [*intra*clasts, *bio*logical remains (for example, shell fragments), *oo*lites, and *pel*lets], the grain size of their cement (*micro*scopic versus *spar*ry appearance), and the overall grain size of their fragments [calc*irudite* (> sand size), calc*arenite* (= sand size), and calc*ilutite* (< sand size)]; see Figure 3–15. As an example, the rather common Floridian rock (often called coquina), which consists of pebble-size shell fragments held together by a submacroscopic cement, could be called either a calcirudite made up of shell fragments and microscopic cement or a biomicrudite, and much of the rather widely used trim stone from Bedford, Indiana, would be a calcarenite with oolites and a microscopic cement or an oomicarenite.

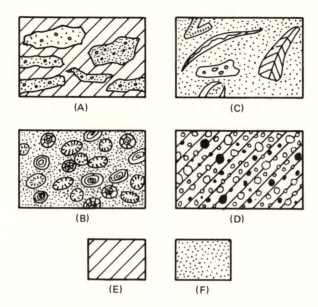

Figure 3–15. (A) Intraclasts in a sparry cement. (B) Oolites (~3×) in microcrystalline (micrite) matrix. (C) Fossils and fragments of fossils in micrite matrix. (D) Pellets (~3×) in sparry cement. (E, and F) Sparry and micritic matrix material respectively. (Reproduced by permission from R. V. Dietrich and B. J. Skinner, *Rocks and Rock Minerals*. © 1979 John Wiley & Sons, Inc., New York.)

REVIEW QUESTIONS

Using Figures 3–14 and 3–15 and Tables 3–2 and 3–3, name the rocks briefly described as follows:

Calcite-cemented _____ the rock consists of approximately 65 percent sand grains, 30 percent pebbles and 5 percent calcite cement.

_____ the rock consists of clay-sized fragments with a preferred orientation giving the rock a fissility.

Hematite-cemented _____ the rock consists of a sand made up of about 35 percent feldspar and 65 percent quartz.

_____ the rock was formed in a cave by the precipitation of the mineral aragonite ($CaCO_3$).

_____ the limestone consists of sand-size oolites cemented together by coarsely crystalline calcite.

Metamorphic rocks: These rocks are generally classified as either *foliated* or *nonfoliated*. The term *foliated* indicates an appearance—for instance, a streaked appearance—that manifests a metamorphically imposed preferred orientation of, for example, platy or tabular mineral grains (Figure 3–6), whereas nonfoliated indicates a rock that has no such macroscopically visible preferred orientation. Common rocks representative of the two categories are shown in Table 3–4.

Table 3–4   Common Metamorphic Rocks.

Name	Features
Foliated	
Gneiss	Imperfect foliation or banding; granular minerals—typically quartz or feldspars—predominate.
Amphibolite	Poorly to well-foliated; green to nearly black amphiboles plus or minus off-white plagioclase predominate.
Schist	Well-developed, closely spaced foliation; platy minerals—commonly one or more of the micas or a chlorite—appear to predominate.
Phyllite	Intermediate between schist and slate; glossy luster; many are corrugated.
Slate	Homogeneous appearing; so fine grained that constituent minerals cannot be distinguished under a hand lens; can be readily split into thin slabs the planes of which need not be parallel to the original bedding.
Typically nonfoliated	
Marble	Typically sparry (that is, fine- to coarse-crystalline); may be predominantly calcite or dolomite.
Quartzite	Conchoidal fracture (see Figure 2–18) is characteristic; typically nearly pure quartz.
Meta-	Prefix used for several metamorphic rocks whose precursors are known—for example, metaconglomerate and metagreywacke.

Two metamorphic rock terms are subject to frequent misuse. The first is *marble*. Geologists apply the term to metamorphic rocks composed predominantly of calcite, dolomite, or both; nongeologists, such as architects, use the term to designate any calcite- or dolomite-rich rock that will take a good polish. This means that several limestones and dolostones, as well as true marbles, are often called "marbles" on the market-place. Unfortunately, specimens whose source or rock associations are unknown are not macroscopically distinguishable as metamorphic as opposed to sedimentary or diagenetic.

The second frequently misused term is *quartzite*. This term is usually used to describe the way rocks that consist largely of sand grains break (Figure 3–16). According to this usage, many silica-cemented sandstones, which are sedimentary, as well as metamorphosed sandstones, are called quartzites. Consequently, if one of these rock's origin is known, it is best to refer to it as either *metamorphic* (or *meta-*) *quartzite* or *sedimentary quartzite*, as the case may be.

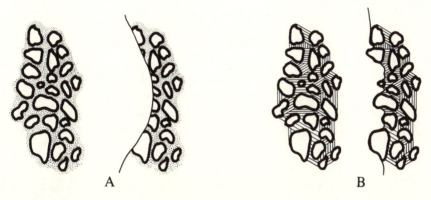

A   B

Figure 3–16. (A) Left: Quartz grains with interstitial quartz cement. Right: Relatively smooth conchoidal fracture cutting indiscriminately across the sand grains and the cement. (B) Left: Quartz grains with interstitial nonquartz cement. Right: Rough surface caused by fracture that is characteristically within the cement and thus around the sand grains. (Reproduced by permission from R. V. Dietrich and B. J. Skinner, *Rocks and Rock Minerals.* © 1979 John Wiley & Sons, Inc., New York.)

REVIEW QUESTIONS

How might you distinguish between a shale and a slate? ⎯⎯⎯⎯⎯⎯⎯⎯

⎯⎯

⎯⎯

If the foliation of a schist were at an angle to the sedimentary layering in its precursor rock, how would you account for the relationship? ⎯⎯⎯⎯⎯⎯

⎯⎯

⎯⎯

In contact (igneous) metamorphism, some rocks have their mineral compositions changed chiefly in response to heating; others have their mineralogical makeups changed because volatiles from the cooling magma move through the rocks, adding and subtracting elements as they go. The former products are called *hornfels;* the latter, *tactites.* What properties of a precursor rock would make it tend to form a hornfels as opposed to a tactite? _____

_____

_____

In the field, you have just found an outcrop of quartzite. What would you look for to determine whether the quartzite is metamorphic or a sedimentary silica-cemented sandstone? _____

_____

Other Rocks: The rocks covered in this section are those already mentioned on pages 60–61.

  Pyroclastic Rocks: Related to both igneous rocks and detrital sedimentary rocks, pyroclastic rocks—all of which may be called *tuffs*— are given names that reflect both relationships. The names consist of four terms. These, in order, are (1) the type of fragment (Figure 3–17); (2) the composition, expressed by the name of

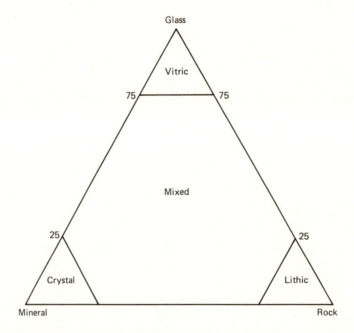

Figure 3–17.   The nature of pyroclastic fragments—glass, mineral, or rock—is frequently indicated by adjectives, as indicated on this triangle. (Reproduced by permission from R. V. Dietrich and B. J. Skinner, *Rocks and Rock Minerals.* © 1979 John Wiley & Sons, Inc., New York.)

the equivalent aphanite (see Table 3–1); (3) the size of the majority of the fragments (Table 3–5); and, in most cases, (4) the word *tuff*, which shows that the rock is a pyroclastic rock (Table 3–5). An example would be *lithic andesite lapilli tuff*.

Table 3–5   Pyroclasts, Tephra, and Pyroclastic Rocks.

Size, Millimeters	Pyroclast (Fragment)	Tephra (Unconsolidated Material)	Pyroclastic Rock (Consolidated Material)
> 64	Bomb*	Bombs	Agglomerate†
	Block**	Blocks	Pyroclastic breccia‡
2§–64	Lapillus	Lapilli	Lapilli tuff
< 2§	Ash grain	Ash	Ash tuff

*Fragment made up of material that was at least partly fluid when ejected.
**Fragment that was solid when ejected.
†*Or* bomb tuff.
‡*Or* block tuff.
§Some geologists use 4 millimeters as the cutoff.
Individual pyroclasts may be described as follows:
   Bombs have twisted shapes that indicate that they solidified from a fluid or partially fluid material while in flight through the air.
   Blocks are large chunks of rock broken from the sides of the volcanic conduit or from a crust covering the underlying magma
   Lapillus (plural lapilli) is the name given to individual fragments of mineral and/or rock (including glass) materials as well as to bomb- and/or blocklike pyroclasts with mean diameters ranging between 2 and 64 millimeters.
   Ash grain is a mineral, glass or rock fragment that is less than 2 millimeters in diameter.
(From Dietrich, R. V., and Skinner, B. J., *Rocks and Rock Minerals*. New York: Wiley, 1979, p. 160.)

Diagenetic Rocks: In many cases, diagenetic rocks are treated as sedimentary. Therefore their names are applied in essentially the same fashion as for precipitated sedimentary rocks; that is, their names reflect their mineralogical compositions. The most common of these are certain limestones (made up wholly or largely of calcite); dolostone (also called *dolomite*), which is made up predominantly of the mineral dolomite; and chert (some of which is often called *flint*), which consists wholly or largely of microscopic to submicroscopic quartz; and the coals. Terms such as *recrystallized limestone* and *dolomitized biomicrudite* are probably preferable. Several other less common rocks also belong to this category.

Migmatites: The nomenclature for migmatites is even more mixed up than the rocks themselves. In fact, these rocks may be better described by illustrations with captions than by words alone. We think it best for all but migmatite experts to call all these rocks simply migmatites and then to note their chief rock constituents; for example, a migmatite consisting of amphibolite and granodiorite.

Veins: These masses are generally named on the basis of their constituent minerals; for example, a galena- and sphalerite-bearing calcite vein (or, because galena often serves as a lead ore and sphalerite as a zinc ore, lead-zinc veins). Especially in the past, some geologists used an alternative terminology indicating conditions thought to have existed during the vein's formation—for example, mesothermal veins. Actually, the naming of the constituent minerals serves a much more useful purpose.

Weathering Products: We include these materials among "other rocks" largely as an example of rocks that are different from the others in their general mode of origin. The classification and identification of these rocks, other than as implied on page 61, are beyond the scope of this book. For information about these rocks, see *Economic Mineral Deposits* by Bateman and *Rocks and Rock Minerals* by Dietrich and Skinner (see references at end of this chapter).

REVIEW QUESTIONS

The proper name for a pyroclastic tuff consisting of ash-size glass fragments of rhyolitic composition would be _____
Could fossils be found in tuffs? _____
If you found a fossil of an animal known to excrete aragonite ($CaCO_3$) exoskeletons now consisting wholly of chert (SiO_2), would you consider the chert a sedimentary precipitate or of diagenetic origin? Why? _____

_____

If you found a composite rock consisting of an intimate intermixture of dark-colored biotite gneiss and granite, what would you call it? _____
If you knew that a pyrite-quartz vein was formed at moderate temperatures, would it be better to call the vein a pyrite-quartz vein or a mesothermal vein? Why?

_____

PSEUDO-ROCKS

There are several man-made materials that are rocklike. Some of them occur along with rocks, especially where rocks are found as loose fragments, as along stony beaches. The most common of these pseudo-rocks are brick, tile, pottery, cinders, coke, concrete, glass, and slag. None of them will be confused with true rocks as long as you are careful not to deal with loose pieces.

REVIEW QUESTION

If you are a collector of rocks, how can you be sure not to include pseudo-rocks?

REFERENCES AND SUGGESTED ADDITIONAL READING

American Geological Institute, *Dictionary of Geological Terms,* rev. ed. Garden City, N.Y.: Anchor Books, 1976.

Bateman, A. M., *Economic Mineral Deposits,* 2nd ed. New York: Wiley, 1950. (This edition is more useful in many ways than the later third edition.)

Dietrich, R. V., *Stones: Their Collection, Identification, and Uses.* San Francisco: W. H. Freeman, 1980.

Dietrich, R. V., and Skinner, B. J., *Rocks and Rock Minerals.* New York: Wiley, 1979.

Flint, R. F., and Skinner, B. J., *Physical Geology,* 2nd ed. New York: Wiley, 1976. (This is one of many introductory geology textbooks that could be consulted in conjunction with the use of this manual.)

Also: The American Association of Petroleum Geologists (P. O. Box 979, Tulsa, Okla. 74101) publishes geological highway maps for different regions of North America.

CHAPTER 4

Fossils

Fossils are useful in many ways. They are useful to scientists because they provide a means of tracing the history of life from its beginnings more than 3.5 billion years ago to the present. They are useful to geologists, in particular, as indicators of how long ago rocks were deposited and what the environment was like at the time.

Fossils are also useful economically. Coal is formed by the compaction of plants that lived in swamps millions of years ago. Oil and gas are the result of chemical changes that took place when buried organisms were subjected to heat and pressure. Many building rocks are fossiliferous limestones. The great pyramids of Egypt, for instance, were built from nummulitic limestone; that is, limestone containing the fossils of nummulities, one-celled organisms that lived about 50 million years ago.

Fossils are also fun to collect and display. Many people derive hours of outdoor enjoyment collecting fossils, and great satisfaction from cleaning, naming, and displaying their collections.

In this chapter you will learn how fossils are named and classified; where they occur; how they are preserved; how to identify a fossil; what fossils can tell you about the past; and where to collect fossils.

REVIEW QUESTIONS

List three ways in which fossils are important to human activities.

1. _____
2. _____
3. _____

Of the things we will cover in this chapter, which topic do you expect to find the most interesting? _____

_____

_____

(Check this answer after you finish the chapter to see if you have changed your mind.)

Have you ever made a fossil collection? If so, what became of it? Are you still actively collecting or do you just have a box of interesting fossils you've picked up over the years? What do you plan to do with the collection? _____

_____

_____

_____

_____

_____

_____

DEFINITION OF FOSSIL

Fossils are the remains of or evidence for plants and animals preserved in rocks. As you can see, for something to be called a fossil, it must:

1. Be the remains of a plant or animal

or

2. Be evidence for the existence of a plant or animal

and

3. Occur in rocks

Each aspect of the definition of a fossil warrants further consideration.

Those plants and animals with the greatest chance of being preserved have some type of hard part—for example, shells, bones, skeletons, or teeth. When an animal dies, its flesh decays and generally only its hard parts are left. If these hard parts are preserved in rock, they become the *remains* of that animal. But even with hard parts, the likelihood of an animal or plant's being fossilized is still not great.

Many animals and most plants, however, have no hard parts. Fortunately, this does not necessarily mean they cannot become fossils. As the definition states, *evidence* of plants or animals also constitutes a fossil. That is to say, such things as impressions of plants or animals, tracks and trails of animals, or even the fossilized feces of animals constitute *evidence* that organisms were once living.

Finally, fossils occur *in rocks*. The vast majority of fossils are found in sedimentary rocks, but under certain conditions, they can be preserved in other kinds of rocks. Ash falls have provided us with many examples of beautifully preserved fossils, such as leaf impressions, insect wings, and fish. Metamorphic rocks in which metamorphism has not been too severe may also contain fossils; for example, fossils have been found in some slates, phyllites, and even schists and gneisses.

REVIEW QUESTIONS

List five examples of fossils.

1. _____
2. _____
3. _____
4. _____
5. _____

Are dinosaur footprints fossils? Why or why not? _____
_____
_____

Would a human footprint on the beach be considered a fossil? Why or why not?
_____
_____
_____

If modern colonial corals live in warm, shallow seas today, and you were to find a fossil colonial coral, what would that fossil indicate about the environment in which the rock was probably deposited? _____
_____
_____

In what type of rock are the majority of fossils preserved? _____
How could a plant or animal be preserved in a pyroclastic rock? _____
_____
_____

THE NAMING AND CLASSIFICATION OF PLANTS AND ANIMALS

Just as humans are unique, so too are the diverse plants and animals of the world. Consequently, we must have some means of naming and classifying plants and animals if we are to make sense of their almost infinite number and to understand the history and evolution of life through time.

A system of classification should group organisms that are alike and separate them from those that are different in such a way that the category highest in the hierarchy contains the most organisms, many of them with only a few things in common, while categories lower in the hierarchy contain members with more characteristics in common. The smallest group, lowest in the hierarchy, would be made up wholly of individuals that are all the same.

Such a system of plant and animal classification, called *taxonomy,* allows us to categorize plants and animals so that we can better study them and their relationships to each other.

In the biological classification, which is used for fossils as well as for living forms, the smallest unit of classification is the *species.* A species is a group of organisms that has the same characteristics and is different from all other species in the world. The presumption is that any two members of the same species can produce offspring that, in turn, are also capable of reproduction, whereas members of two different species, no matter how closely related, cannot produce such offspring. To make the nomenclature system useful, each species must have a name unique to it, thus setting it apart from all other species. Such a system would be extremely unwieldly if each species were to have just one name; with over 1,600,000 known plant and animal species, we would soon run out of names! Thus biologists and paleontologists use the system of *binomial* (that is, two-name) *nomenclature,* introduced in 1758 by the Swedish naturalist, Carl von Linne, and now used on a worldwide basis by all scientists to describe and name both living and fossilized plants and animals.

This binomial system requires that each species' name consist of a *generic* name (the name of the *genus* to which the species belongs) and a *specific* name (the name of the *species*). In print, both of these names are italicized, and the first letter of the genus name is capitalized whereas all other letters are lowercase. Where the genus and specific names cannot be put in italics, they are underlined. Both names are latinized and generally come from either Greek or Latin. The specific name is usually a descriptive term for the organism, although it can be the name of a person or a place.

The organic world can be divided into two major divisions, plants and animals, called the *plant kingdom* and *animal kingdom* respectively. The kingdoms can be divided further into *phyla* (*phylum,* singular); the phyla divided into *classes;* classes into *orders;* orders into *families;* families into *genera* (*genus,* singular); and genera into *species.* Furthermore, each of these categories may be subdivided into subgroups such as *suborders* or *subspecies,* or grouped together in higher groups such as *superfamilies* and *supergenera.* The *super* and *sub* groupings, however, are not often used.

Thus, we see that our classification is a hierarchical system in which the members of each lower group have more characteristics in common than the members of the next higher group. A comparison of the classification of humans and dogs illustrates this system (Table 4–1). Humans and dogs both belong to the animal kingdom. They both belong to the phylum Chordata, which includes all animals with a spinal cord, and they both also belong to the class Mammalia, which includes all animals that are warm blooded and have mammary glands and hair. Therefore, to this point, dogs and humans are in the same categories, since both have characteristics that place them in these large categories. At the level of order, however, the significant characteristics are specific enough to differentiate humans and dogs. Dogs belong in the order Carnivora, which also includes cats, whereas humans belong in the order Primates, which also includes apes and monkeys. At this point, dogs and humans no longer have characteristics in common and at the family and genus level are quite far removed from each other. At the species level, all individuals within the species have the same general biological characteristics. It is obvious that there are differences among humans and among dogs. These differences, however, are merely variations in the same characteristics, such as height, weight, or length and shape of noses, and so on.

Table 4–1   Classification Comparing Humans and Dogs.

Taxonomic Unit	Humans	Dogs
Kingdom	Animalia	Animalia
Phylum	Chordata	Chordata
Class	Mammalia	Mammalia
Order	Primates	Carnivora
Family	Hominidae	Canidae
Genus	*Homo*	*Canis*
Species	*Homo sapiens*	*Canis familaris*

You will frequently see the name of a person and a date after a species name. This refers to the person who first named the organism and the date when it was formally named. For example, *Athypella carinata* Johnson, 1964 is a brachiopod that was named and described by Johnson in 1964. You may also see a name in parentheses following a species name. This means that the species has been transferred to a genus different from the one in which it was originally placed by the person who named it. For example, the name *Mucrospirifer alpenesis* (Grabau) Stumm, 1956 means that Grabau originally named the species and Stumm transferred it to the genus *Mucrospirifer* in 1956.

Before learning how to use the appropriate taxonomy for identifying fossils, let's see how well you understand how living things are named and classified.

REVIEW QUESTIONS

What is one of the purposes of a classification system for plants and animals?

_____

_____

_____

Define *taxonomy* and state its purpose. _____

_____

Why do biologists and paleontologists use two-part names for organisms? ____

_____

_____

What is the order of classification of organisms, from highest category to lowest category in the hierarchy? _____

_____

_____

_____

_____

_____

What is the meaning of a person's name and a date after an organism's scientific name? What does it mean if a person's name is in parentheses? _____

_____

_____

_____

_____

USING TAXONOMY TO IDENTIFY FOSSILS

As we have seen, plants and animals are grouped into categories that reflect their similarities. When identifying a fossil, the collector should first determine the phylum and class to which a specimen belongs. This is generally not too difficult, because there are not many groups to choose from. Except for an extremely unusual specimen, most collectors soon learn how to place a specimen in its proper phylum and class. From that point, it is a matter of identifying the specimen down to its correct genus and species. This is usually done with the aid of a specialized book on fossils or a local guidebook that lists the common fossils occurring in a particular area. Appendix V will enable you to identify most, if not all, of your fossils at least down to their order. The characteristics of each invertebrate phylum are listed along with their morphology (that is, their shape and form) and an illustration of some of the more common species. A list of some good fossil books, as well as sources of other useful publications, is given at the end of this chapter.

After you have collected some fossils, how do you go about identifying them? The first step is to obtain a book or books that identify the fossils from the area where you collected (Figure 4–1). These may be state or federal publications or scientific periodicals. Such publications are available in most university libraries and also in some larger public libraries.

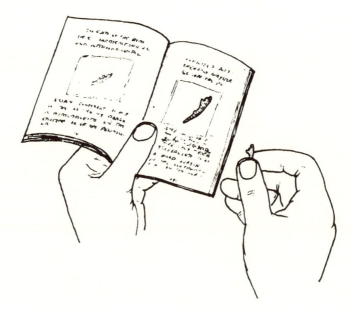

Figure 4–1.   Identifying a fossil.

Before trying to name your fossils, group all members of the same phylum together, then group members of each phylum into smaller categories so that all the fossils in each category are essentially the same, and thus probably belong to the same species. Obvious morphological differences, such as the number of ribs on a brachiopod or the tooth-and-socket arrangement of a pelecypod, are the kind of characteristics that should be used to separate species from each other within a phylum. Other characteristics, such as size and color, are generally not important in separating most species. With practice, and the use of fossil identification books for the areas you are collecting, you can begin to get an idea of what differences are important as well as the amount of variation you may expect for each species.

If you know to what phylum your specimens belong, go to that section of your book and compare your specimens with the illustrations there. When you find one or several illustrations that appear similar to your specimens, read the descriptions carefully and note the class, order, family, and genus to which each belongs. You should then be able to determine the class, order, and probably the family of your specimens. Remember, you are always going from the general to the specific when classifying organisms, so, in the end, you want to classify your specimens as precisely as possible. Therefore, even though you may not be able to identify your specimens down to their species, be sure to identify them down to the lowest level of classification that you can. Many books for a specific area, however, give only the characteristics of the various species to be found. In this case, you will be identifying your fossils to their species level immediately.

As you become more proficient at identification, you will become familiar with the various books and journals that are useful to you, and you will be able to identify almost all your specimens down to the species level. It is only with practice and familiarity with the paleontologic literature that you will become truly proficient at identifying fossils. Eventually, you will be able to recognize many different species on sight. You may even be surprised at how many species you do know!

REVIEW QUESTIONS

What is the first thing you should do after collecting your fossils and before trying to identify them? _____
_____

How are plants and animals classified? _____
_____

Where can you find books and journals useful in fossil identification? _____
_____

Have you ever tried to identify fossils before? Were you successful? If not, why not? _____
_____
_____
_____

How do you use a fossil guidebook or article about fossils in your area to help you identify the fossils you find? _____
_____
_____
_____

THE FORMATION OF FOSSILS

We have seen that fossils are the remains of or evidence for plants and animals preserved in rocks. The organisms with the best chance of being fossilized are those with hard parts that, after death, are covered rapidly by sediments. Thus, the majority of fossils are found in marine sedimentary rocks. Only a very small percentage of living plants or animals are ever preserved as fossils. Approximately 1,500,000 species of living plants and animals have been described and named, whereas only about 130,000 species of plants and animals are known as fossils. This discrepancy becomes even more striking when one realizes that we are comparing those living things in existence during an instant of geologic time, the present, with those from all of the geologic past since life began, approximately 3.5 billion years ago.

Actually, if preservation was only reasonably good, one would expect to find both more fossils and more species as fossils than have been discovered. Why then do we not find more fossils and more species as fossils? One reason is that most plants and animals die in environments that are not suitable for fossilization. Another is that, regardless of the environment in which an organism lives, there are many additional factors working against its being fossilized. First of all, as soon as an organism dies, it is attacked by predators and scavengers that not only will devour an animal's flesh, but also will scatter its bones. At this stage, whatever is left will be subject to bacterial degradation, which speeds up the decay process. Second, the hard parts of the organism, if any, may be subject to mechanical destruction, such as that imposed during movement of those hard parts from place to place and resulting from abrasion by sedimentary particles. This can occur either on land or in water. Third, even if an organism survives destruction, and is buried, chemical destruction by solution (dissolving), may still prevent the organism from becoming a fossil. Consequently, it is fortunate that *any* once-living plants and animals became fossils.

Obviously, organisms that possess hard parts and are buried rapidly have the best chance, however slight, of being fossilized (Figure 4–2). From what types of material do organisms produce shells, bone, teeth, and other hard parts that are potentially fossilizable? In spite of the great diversity of life, only a few substances are used for hard parts. In order of abundance, they are as follows:

1. *Calcium carbonate,* $CaCO_3$. Calcite and aragonite are two common minerals that have the same chemical formula, $CaCO_3$, but a different structural arrangement of their constituent atoms. These constituents are present in seawater, and a majority of marine invertebrates use them in shell construction. Because aragonite weathers and becomes altered more readily than calcite, fossils of invertebrate animals that use aragonite are generally not as well preserved as those that use calcite. Calcite is used by species belonging to 13 of the invertebrate phyla, while aragonite is used by species belonging to four invertebrate phyla.

2. *Silica,* SiO_2. Silica is used only by members of three different groups: diatoms, radiolarians, and a few sponges. It is extremely resistant to decay.

3. *Calcium phosphate,* $Ca_5(PO_4)_3OH$. This is the compound that inarticulate brachiopods and trilobites use for construction of their shells. It is also the compound of which bones and teeth are composed.

4. *Organic material.* This category includes several rather complex organic compounds, such as cellulose, cutin, lignin, and other materials used especially by plants. It is usually not preserved in the rock record except as carbonaceous films that represent impressions.

Figure 4–2. How an animal becomes a fossil. (A) Living dinosaur, 185 million years ago. (B) Dinosaur dies. (C) Area is covered by water and sediments. (D) Dinosaur is buried in sediment. (E) Fossil collector digging up remains of dinosaur bones.

Before proceeding to a discussion of the various methods of fossil preservation, let's see how much you have already learned about the general subject.

REVIEW QUESTIONS

Which organisms have the best chance of being fossils? Why? _____

_____

Are there more living plant and animal species or fossil plant and animal species? Why? _____

_____

List three things that tend to prevent an organism from becoming a fossil.

    1. _____
    2. _____
    3. _____

Why are there more fossil animal species than fossil plant species? _____

_____

What are the four most common materials of which the hard parts of plants and animals are made?

    1. _____
    2. _____
    3. _____
    4. _____

KINDS OF PRESERVATION

There are several ways in which organisms may be preserved as fossils.

Preservation Without Alteration: Fossilization in which the original material is preserved without any apparent alteration is the most common means of preservation of marine invertebrates (animals without backbones). Remember, those organisms with the best chance of being fossilized have hard parts and are buried rapidly. These are the conditions that exist for invertebrates living and dying within the oceans. The oceans, teeming with life that has hard parts, and receiving sediments from the land, offer the best chance for fossilization to occur. Thus, marine organisms have the best chance of being preserved, and their shells are generally preserved without any visible alteration.

But what about land plants and animals? How are they preserved? As you might guess, some of them can be fossilized just like their marine counterparts; all that is required are conditions suitable for such preservation. These conditions are found in some lakes, and even in a few streams. For the majority of land plants and animals, however, the likelihood of becoming fossilized is very small. Nonetheless, there are three special kinds of preservation of land animals and plants that are of interest, though rare.

The first of these is *amberization,* which provides us with information about insect and plant life of the past (Figure 4–3). When the bark of certain trees, particularly some conifers, is ruptured, a thick, sticky resin flows out. This resin provides a trap for many crawling and flying insects, and also for windblown pollen, seeds, twigs, and so on. The insect, for example, becomes stuck in the resin and eventually may become embedded in it as more resin flows around it. Under suitable conditions, this resin may itself be preserved in sediment and eventually be converted to amber.

Figure 4–3. Amberization. Caddis fly preserved in amber. Oligocene of East Prussia.

Perhaps the best-known amber deposits are in the Baltic region of Europe. Insects that lived during the Eocene epoch, 50 million years ago, are preserved in these amber deposits. The degree of preservation is so remarkable that delicate features, such as hairs, wings, antennae, and muscle tissue, have been preserved; in fact, there is even a report of fossil spider silk within amber. Such amberization has allowed paleontologists to study the insect world of the geologic past and to follow the evolutionary history of several of the insect lineages.

A word of caution is in order for the novice collector. Some unscrupulous dealers sell what they claim to be fossil insects preserved in amber. Upon initial examination, these "fossils" look genuine, but careful examination will reveal that they are "too perfect." Many have been manufactured by humans; it is easy to heat amber, drop insects into it, and then let it cool. The key to recognizing such a fraud is the "too perfect" quality of the "fossil" and the enclosing amber, which is perfectly clean with no dirt or air bubbles. In the real world, an insect will struggle with all its strength to break free of the sticky resin in which it is trapped. In many cases, this causes air bubbles and streaks, as well as dirt or bits of plant debris, to also be entrapped. Much amber, therefore, is not perfectly clean and neat, but rather dirty. Thus, before buying what is purported to be fossil amber, check it carefully and, if it looks too perfect, it may be prudent to pass up the opportunity to purchase it.

A second unusual, but interesting, method of preservation without alteration is *impregnation by tar or asphalt*. Perhaps the best-known example of this is the La Brea Tar Pits in southern California (Figure 4–4). These tar pits, existing since the Pleistocene Epoch, have provided paleontologists with an amazing record of life in a semiarid area during the time when glaciers covered the northern portions of the continent. In the Rancho La Brea area of southern California, cracks and fractures in the earth's crust provided the pathways for heavy crude oil to migrate to the surface, where it bubbled out onto the ground into small lakes and watering holes. Many animals and birds came to these lakes and watering holes to drink. Some fell into the water and got stuck in the oil. As they struggled to escape, they attracted predators and scavengers that also became stuck in the heavy oil. The prey and the predators moored in the gooey tar and asphalt subsequently sank to the bottom of the pools, where their flesh rotted and their bones became impregnated with the entombing tar. Thus, a fairly representative cross section of the population of the area was preserved, including deer, elephants, bison, camels, giant sloths, horses,

Figure 4–4.   Impregnation by asphalt. Mammoth caught in asphalt at Rancho La Brea, southern California.

lynxes, leopards, wolves, saber-tooth tigers, birds, and insects. Many of these animals became extinct about 10,000 years ago, and without this rare type of fossilization it is unlikely that we would have as complete a record as we do for this kind of area for the Pleistocene.

Freezing in ice, the third and rarest method of preservation, has provided us with skin, hair, and even the stomach contents of extinct woolly mammoths that roamed the frozen Siberian and Alaskan plains 10,000 to 100,000 years ago. These large beasts apparently either fell into supercooled lakes, where the water instantly froze around them, or fell into crevasses and were subsequently covered by ice. In one instance, a mammoth was frozen so quickly that its last mouthful of food was found unchewed in its mouth. The first reported occurrence of a frozen mammoth was from the Lena River Delta of Siberia, in 1799. Since then, more than 50 additional discoveries have been made. The flesh is so well preserved that it has been eaten by wild animals and sled dogs. One of the most exciting finds was that of a perfectly preserved 44,000-year-old baby mammoth in northeastern Siberia (Figure 4–5). It was so well preserved that some of its red blood cells were still intact, the first known example of fossil blood cells. An attempt currently is underway to analyze its genetic makeup and amino-acid sequence.

Figure 4–5.   Freezing in ice. Juvenile mammoth (44,000 years old) preserved in permafrost, from northeastern Siberia, U.S.S.R.

Before studying other methods of preservation, let us see what you have learned about preservation without alteration.

REVIEW QUESTIONS

What kinds of organisms have the best chance of being preserved? Why? _____
_____
_____
_____

What kind of environment is most suitable for fossilization? Why? _____
_____
_____

What kind of environment is the least suitable for fossilization? Why? _____
_____
_____

What is amberization? How does it occur? _____
_____
_____
_____

What organisms (plants or animals) are preserved in amber? How? _____
_____
_____
_____

How can you generally recognize man-made "amber fossils"? _____
_____
_____
_____

How are animals preserved in tar? _____
_____
_____

Where are the best places to find examples of amberization, impregnation by tar, and freezing? _____
_____
_____
_____

What can freezing tell us about an animal that impregnation by tar cannot? ___
_____
_____
_____
_____

Preservation with Alteration: There are several methods of preservation whereby the original material of the organism is altered. These methods often increase the chances for fossilization. The first type of alteration is a group of processes usually termed *petrification*. The word comes from the Greek *petros,* which means "stone," and petrification thus means "turned to stone." While many people consider any fossil to represent petrification, strictly speaking an organism can be petrified only if secondary material has been deposited in its pores or cavities, or its original material has been replaced. This can occur in three ways, each of which is distinct from the others and has a distinctive name.

Permineralization occurs when there is secondary mineralization of the original organism's hard parts. It is usually accomplished by the filling-in of the pores or empty spaces in a shell or bone. Quartz is the most common mineral involved. Permineralization preserves the gross structure of an organism but not necessarily its detailed structure. It has the beneficial effect of preserving many organisms that ordinarily would not be preserved. In addition, it usually results in the hardening of the preserved organism. Petrified wood such as that in the Petrified Forest in Arizona, is an excellent example of permineralization (Figure 4–6).

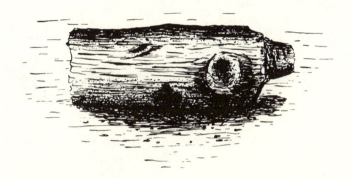

Figure 4–6. Permineralization. Petrified log from the Petrified Forest, Arizona.

Replacement, also termed *histometabasis,* is volume-for-volume replacement of the original material by a secondary mineral, so that the fine cellular structure is preserved in great detail. As in permineralization, the secondary minerals are frequently derived from saturated groundwater as they circulate through the entombing sediments.

More than 50 minerals have been known to replace original organic material, but the most common is quartz. The replacement of calcite by quartz in Permian-aged brachiopods from the Glass Mountains in Texas has provided us with especially beautifully preserved external structures of those brachiopods (Figure 4–7A). The Miocene-aged silicified insects from the Calico Mountains of southern California, however, are perhaps the most spectacular examples of replacement known (Figure 4–7B). These insects occur in calcite-rich concretions. When the concretions are dissolved in dilute hydrochloric acid, the insects are all that remain. The original organic exoskeleton has been replaced by silica, which exquisitely preserves all the details of the insects. This unusual deposit has enabled paleontologists to learn much about the insect fauna of the Miocene Epoch.

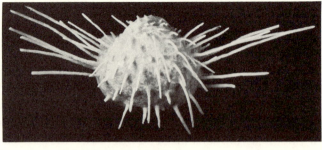

A

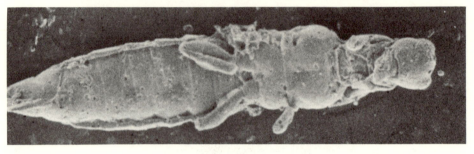

B

Figure 4–7. Replacement. (A) Brachiopod *Echinauris,* Glass Mountains, Texas. (Photograph courtesy of Smithsonian Institution.) (B) Miocene insect, Calico Mountains, southern California.

A few of the other relatively common replacing minerals are calcite, dolomite, pyrite, hematite, limonite, and glauconite.

The chief difference between permineralization and replacement is that, in the former, only the gross structure is preserved, whereas in the latter, the individual detail is commonly preserved.

The third type of pertrification is *pseudomorphism* (literally, "false formism"). This occurs when the original material or organism rots away or is otherwise removed, leaving a mold that is then filled by mineral matter, commonly quartz. The result is a fossil that looks like the organism on the outside, but on the inside exhibits no organic structures. Pseudomorphic plant forms have been found in lava flows in the Pacific Northwest where the plants were engulfed, covered by lava, and burned away, leaving a mold of their external shape. The fossil was formed when the mold was later filled by quartz.

Still another type of preservation with alteration, which does not involve petrification, is *carbonization*. Carbonization results when soft-bodied animals or plants are buried in fine-grained sediments or pyroclastics. The volatile components resulting from the decay of the soft parts are driven off during burial, thus leaving only a fine carbonaceous film representing the original organism (Figure 4–8). This is the most common method of preservation for plants, and also provides us with information about soft-bodied organisms such as jellyfish and worms, which are, in general, not otherwise preserved because of their lack of hard parts. An excellent example of this fossilization process is the beautifully preserved Middle Cambrian soft-bodied animals from the Burgess Shale of British Columbia, Canada.

Figure 4–8. Carbonization. Frond impression from Bluefield District, Virginia. (Photograph courtesy of Norfolk and Western Railway.)

Another type of preservation with alteration is called *recrystallization*. This results when shell material is subjected to conditions such that its crystal structure is altered. This recrystallization usually obliterates the fine detail of the organisms so affected; it occurs most frequently when limestone is converted to dolostone or is recrystallized to diagenetic limestone.

Let us check now to see what you have learned about preservation involving alteration.

REVIEW QUESTIONS

What does the word *petrification* mean? _____

How does permineralization differ from replacement? Give an example of each.

_____

_____

What does *histometabasis* mean? _____

_____

What is the most common secondary mineral in replacement? In permineralization? _____

_____

Name three other secondary replacing minerals.

    1. _____

    2. _____

    3. _____

What is a pseudomorphic fossil? How is it formed? _____

_____

_____

Using wood as an example, briefly describe how it could be permineralized, replaced, or made into a pseudomorphic fossil. Tell how you would determine which method of alteration took place. _____

_____

_____

_____

_____

What is carbonization? What type of living thing is most likely to be carbonized?

_____

_____

What is recrystallization? What type of rock is most susceptible to recrystallization? What is a common resulting rock? _____

_____

_____

EVIDENCE OF ORGANISMS

Remember that a fossil can be the evidence for as well as the remains of a plant or animal. We have discussed the different ways we may find an organism preserved; now we shall describe some of the ways one may find evidence of an organism without having any of the original organism preserved or replaced.

A *mold* is the impression a plant or animal leaves in a rock when it is dissolved away or otherwise removed (Figure 4–9). If the impression is of the outside of the plant or animal, it is referred to as an *external mold*. If it reflects the morphology of the insides of the organism, it is called an *internal mold*.

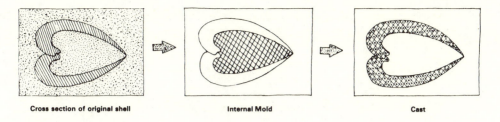

| Cross section of original shell | Internal Mold | Cast |

Figure 4–9. Mold and cast of a pelecypod shell.

A *cast* results when a mold is filled (Figure 4–9), and provides a positive, three-dimensional representation of the original structure. Some casts are natural, but others are man-made, as when a mold is filled by a substance such as plaster of Paris or latex rubber. Making artificial casts is a real skill, which has been highly developed by many collectors, paleontologists, and museum curators.

Molds and casts are encountered rather commonly by collectors of fossils of invertebrate animals. An especially exciting new area of research involves their use in studying the braincases of certain vertebrates. These braincases, which are natural molds, are filled either naturally by sediment or artificially with latex rubber. A cast of the brain is then preserved. From the study of such brain casts, one can plot such things as the increase in brain size for various extinct animals and then attempt to relate this to intelligence. Or we can use the brain casts to help analyze which areas of the brain were dominant and when they became dominant in the geologic past. It is known that specific areas of the brain control specific functions, such as speech and eye–hand coordination; thus, by determining when a given area became dominant or enlarged, we can determine when such functions as speech arose in humans or whether pterosaurs (flying reptiles) were warm-blooded.

Unlike molds and casts, some types of fossil evidence do not directly represent any part of the original organism. For example, many animals leave evidence of their existence in the form of *tracks, trails, burrows,* or *borings.* Such evidence can tell us a lot about the ways the animal moved around and perhaps also about its feeding habits.

Tracks are the footprints made by animals as they walk over soft sediment (Figure 4–10C). Tracks can tell us the way an animal walked, how fast it moved, the length of its limbs, how tall it was, and even approximately how much it weighed. Perhaps one of the most famous sites of fossil footprints is the Triassic beds of the Connecticut Valley. The footprints of several different species of dinosaurs are preserved in these rocks. Interestingly, no dinosaur bones have ever been found in any of these beds! Fossil footprints have also been used to date the time when our human ancestors first walked upright. These human footprints have been found in an ash deposit in Tanzania, Africa, radiometrically dated at 3.5 to 3.8 million years ago.

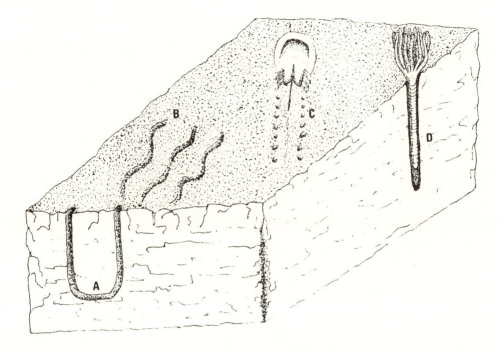

Figure 4–10.   Tracks, trails, and burrows. (A) U-shaped dwelling burrow. (B) Trail left by crawling animal. (C) Tracks left by horseshoe crab. (D) Feeding burrow with worm still in it.

Trails are the impressions made by animals that crawl (Figure 4–10B). Trails, like tracks, often tell us something about the way an animal moved or its feeding pattern.

Burrows are traces made by animals as they move through sediments (Figures 4–10A,D). Burrows can be either *feeding burrows,* which result from an animal moving through sediments in search of food, or *dwelling burrows,* which are tunnels excavated for use as living quarters. Most dwelling burrows are at nearly right angles to the bedding plane.

Whereas most tracks can be assigned to a particular species or genus, most trails and burrows cannot. This is so because many animals make essentially the same kinds of trails or burrows. In most cases, it is only when the animal that made the trails or burrows is preserved near them that a definite assignment can be made.

Borings are holes made by animals into rocks or other solid substances for shelter, or into other animals in search of food. For example, many clams bore into rock and attach themselves inside the holes, often spending the rest of their lives there. And some animals, such as carnivorous snails, bore small perfectly round holes through clam shells, inject a toxin that causes the muscles that hold the shells together to relax, and when the shells open up, then they eat. Such holes in ancient clam-shell fossils tell us that predatory snails were part of the living community, even though their shells have not been preserved.

Such tracks, trails, burrows, and borings are referred to as *trace fossils,* which can be very helpful in interpreting life activities of the past.

Another type of evidence of animal activity is *coprolites,* fossilized feces of animals. These too can provide us with much information about the animal responsible for them. Many coprolites have been found in association with the animals that excreted them; thus, they furnish valuable information about the dietary habits of the animal. Also, even when the coprolites are not found near the parent animal, their size, shape, and external markings can tell us something about the size of the animal and any unusual characteristics of its alimentary tract.

Another type of evidence of animal activity is *gastroliths,* highly polished stones found in the stomach regions of dinosaurs and of some extinct marine reptiles. They are thought to have aided in the grinding of food within the gastrointestinal tract of the animal. Just as naive collectors can be tricked into buying artificially produced amber fossils, they also may be tempted to buy man-polished stones as "gastroliths." It is very difficult to detect man-made gastroliths and hence one usually must rely on the integrity of the dealer.

REVIEW QUESTIONS

How are molds and casts formed? _____
_____
_____

What can molds and casts tell us about animals or plants that the actual remains of the organisms cannot? _____
_____
_____

What is the difference between tracks and trails? _____
_____
_____

What is the difference between burrows and borings? _____

_____

_____

What is the general term for evidence of animal activity? _____
What are coprolites? What can they tell us about the animal that made them?

_____

_____

_____

_____

What are gastroliths? How did they serve their host animal? _____

_____

_____

Do you own any gastroliths? How can you be sure they really are gastroliths?

_____

_____

_____

PSEUDOFOSSILS

Remember that a fossil must be either the remains of or evidence for organic activity. *Pseudofossils* are inorganic objects that bear a superficial resemblance to things of organic origin. Among the more common pseudofossils are the following.

1. *Dendrites*. These are dark branching patterns that occur sporadically on the surfaces of several different rocks. They superficially resemble ferns, and are commonly mistaken for fossil plants (see Figure 2–12). They are, in reality, mineral deposits of some manganese oxide or an iron oxide. They can usually be differentiated from true plant fossils by the fact that they are small, much smaller than true ferns. In addition, dendrites occur in igneous and metamorphic rocks as well as in sedimentary rocks.

2. *Glacial striations and slickensides*. These are, respectively, grooves produced by glaciers moving over a rock, and striations produced when two rock units move past each other along a fault. While these structures may superficially resemble trails, or even burrows, careful examination will reveal that they are all generally parallel and straight, which would be quite rare if produced by animals.

3. *Concretions and weathering products*. Many concretions and weathering products resemble fossils. Close examination will reveal that they lack any structure and form, and are the result of inorganic forces rather than the remains of animals (Figure 4–11).

Figure 4–11.   Pseudofossils: Upper row, left to right: Cone-in-cone, four concretions. Lower left: Septerian concretion. Center: Erosion form. Middle and bottom, right row: Concretions. Right center: Barite "roses." (Photograph courtesy of G. K. McCauley.)

4.   *Vertical tubes*. In some sedimentary rocks, there are vertical tubes that look like burrows made by worms. Some of these tubes may be the result of the escape of gas bubbles through the sediment while it was changing into sedimentary rock. Some of these gas-escape tubes are extremely difficult to distinguish from burrows, and they may fool even professional paleontologists.

5.   *Fossil raindrop prints, fossil ripplemarks, and fossil mudcracks.* You may read about these "fossils" occasionally. Since all are formed by inorganic, though natural, phenomena, the use of the word *fossil* in terms such as these is inaccurate; *fossil* refers only to organically produced features.

REVIEW QUESTIONS

How does a fossil differ from a pseudofossil? _____
_____
_____

How would you distinguish a dendrite from a fossil fern? _____
_____
_____

Do you have any rocks that look like, but are not, fossils? If so, how did you determine they were not fossils? _____
_____
_____

Briefly discuss how inorganically produced "trails" and "borings" are formed.
_____
_____
_____

What is inaccurate about the term *fossil raindrop*? What would be a better name for it? _____
_____
_____
_____

USES FOR FOSSILS

We have already mentioned many uses of fossils. They provide us with a means of tracing the history of life; they can tell us how old a rock is and what the environment was like when the enclosing sediment was deposited; they allow us to correlate rocks from widely separated parts of the world; they are useful in several economic endeavors. Of these uses, the first—the use of fossils in the tracing of the history of life—is perhaps the most important. As we have seen, we can trace the history of life by studying fossils from diverse sedimentary rock units of different ages.

Stratigraphy is the study of stratified rocks. Stratigraphers are concerned with the various characteristics of rock units and how rock units are related to one another. One of the important principles of stratigraphy is the *law of superposition*. It states that, in a flat-lying sequence of rocks, the oldest rocks are on the bottom, and the youngest rocks are on the top. This principle is very important to stratigraphers, as it allows them to place rock units, even those that have been folded, in a chronological order.

Since fossils are found in sedimentary rocks, a corollary to the law of superposition can be formulated. This is the *law of faunal* and *floral succession*, which states that assemblages of fossil plants and animals succeed each other in definite, determinable sequences. Furthermore the presence of similar assemblages in different rock units suggests that the rocks are of the same age. This is true because many fossil assemblages are distinctive for particular time periods. Therefore, a paleontologist who finds the same assemblage of fossils in widely separated areas can conclude that the rock units are of the same age (Figure 4–12).

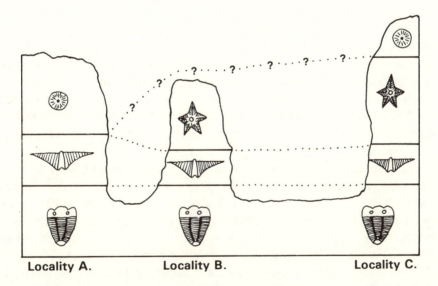

Locality A.   Locality B.   Locality C.

Figure 4–12. Correlation of strata based on fossil content.

Some fossils can also tell us what an environment was like in the geologic past. For example, plants are sensitive to changes in temperature and other factors: some plants, such as cacti, grow primarily in hot, dry areas, while others, such as ferns, grow only in moist areas. Thus, when we find fossils of a particular type of plant and we know its present-day ecological niche, we can infer from the fossil evidence what the environment was like where and when that type of plant lived in the past.

We can do the same thing with some animals. Many marine invertebrate animals are restricted to certain water depths and temperatures. Thus, when we find fossils of them, we can infer what the conditions were like where and when the animal was alive. For example, all present-day colonial corals live in warm, shallow, clear waters; thus, when we find the same kinds of coral as fossils in sedimentary rock, we are reasonably confident that when the sediments that entombed those corals were deposited, they were deposited in warm, shallow, clear water.

Fossils also provide us with evidence that supports the theory of organic evolution. This theory states that all present-day species evolved from earlier species, and that there has been, in general, a progression from simpler, primitive life forms to more complex, advanced forms.

If we look at the fossil record, we see that the oldest rocks contain plants and animals that are much different from the forms we have today, and that, the younger the rocks are, the more similar the fossils are to their modern counterparts. The geologic epochs into which the Cenozoic Era is divided are, in fact, based on the percentage of extinct molluscan species that rocks of each of the periods contain. For example, rocks of the Eocene Epoch contain molluscs approximately 95 percent of which are extinct, whereas rocks of the Pliocene Epoch contain molluscan species approximately 90 percent of which are still alive today.

We can also trace the evolutionary history of organisms through time by studying changes exhibited by the fossil record. One of the best documented records of such evolutionary change is that of the horse (Figure 4–13). North America has one of the most complete sequences of nonmarine Cenozoic Age sedimentary rocks, and the evolutionary history of the horse is recorded within these rocks. Fossils of the earliest horse are present in Eocene Age rocks; they reveal a horse about the size of a fox, with small, simple teeth; a high, arched back; back feet with three toes; and front feet with four toes. Changes that take place in the horse through the Oligocene, Miocene, Pliocene, and Pleistocene Epochs, up to the present-day horse are also present (Figure 4-13).

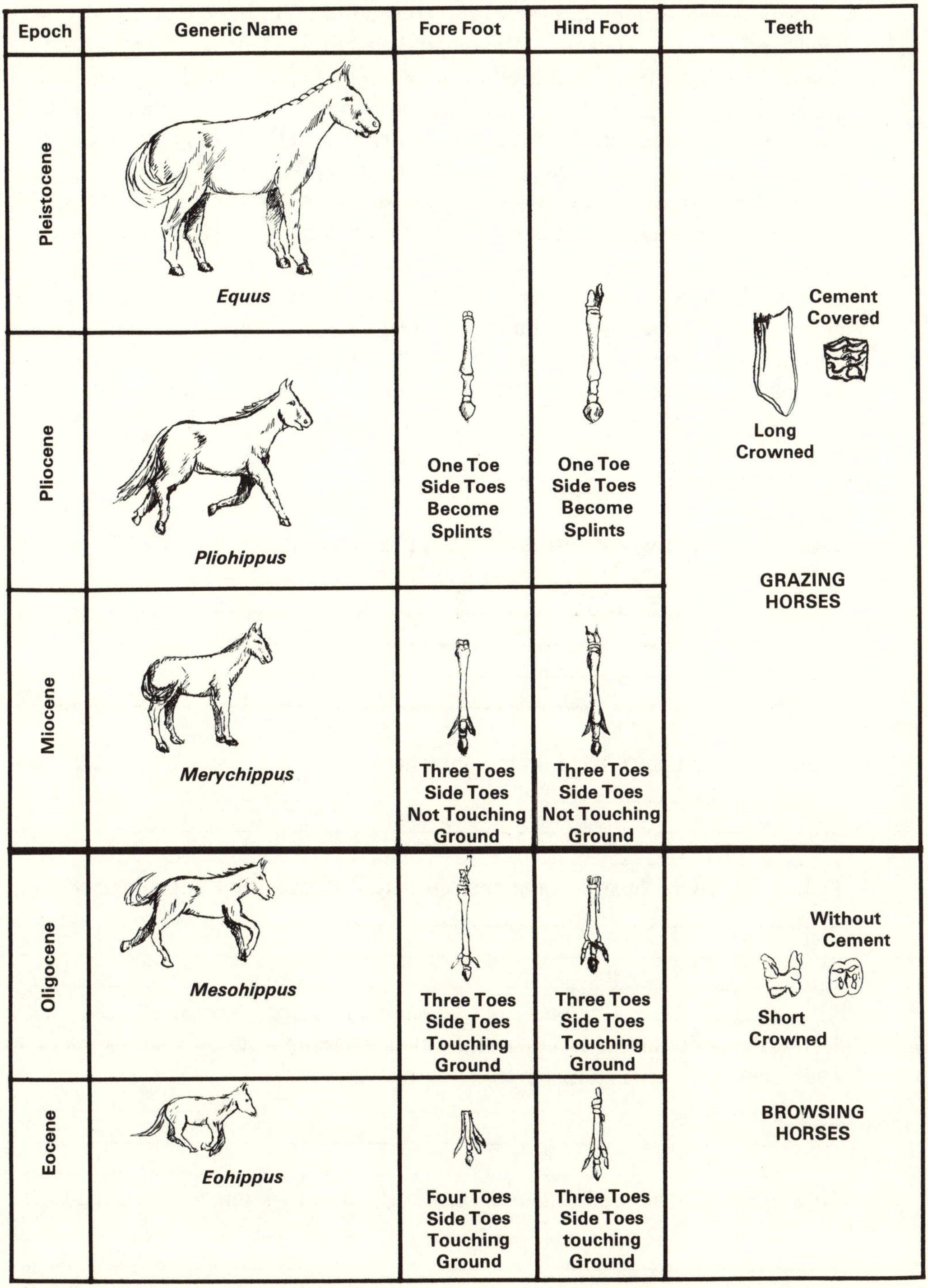

Epoch	Generic Name	Fore Foot	Hind Foot	Teeth
Pleistocene	*Equus*			
Pliocene	*Pliohippus*	One Toe Side Toes Become Splints	One Toe Side Toes Become Splints	Cement Covered / Long Crowned / GRAZING HORSES
Miocene	*Merychippus*	Three Toes Side Toes Not Touching Ground	Three Toes Side Toes Not Touching Ground	
Oligocene	*Mesohippus*	Three Toes Side Toes Touching Ground	Three Toes Side Toes Touching Ground	Without Cement / Short Crowned / BROWSING HORSES
Eocene	*Eohippus*	Four Toes Side Toes Touching Ground	Three Toes Side Toes touching Ground	

Figure 4–13. Evolution of the horse.

The importance of fossils to economics is manifold. The buried remains of plants have been converted into coal. The oil produced today is the result of the conversion of literally trillions and trillions of one-celled plants and animals and their organic matter into petroleum. The discovery and eventual exploitation of these resources are dependent, in part, on the ability to correlate rocks of the same age. This is done, as we have already seen, by the use of fossils. Thus, the use of fossils in stratigraphy provides the petroleum and several other mineral industries with knowledge that saves hundreds of millions of dollars annually in prospecting ventures.

Many fossiliferous limestones are used as building stones. In addition to the numulitic limestones used to build the Great Pyramids, other fossiliferous limestones, such as the Salem Limestone which is quarried principally near Bedford, Indiana, are in great demand as accent stone for buildings and monuments throughout the world.

REVIEW QUESTIONS

List four ways in which fossils are useful to humans.

1. _____
2. _____
3. _____
4. _____

What is stratigraphy? What do stratigraphers do? _____
_____
_____

Define the law of superposition and tell why it is important to geologists. ____
_____
_____
_____

Define the law of faunal succession and give an example of how it can be useful to geologists. _____
_____
_____

How can fossils be used to determine the age of a rock unit? _____
_____
_____

How can fossils tell us what the environment of the earth was like at a particular time in the past? _____

_____

_____

How can fossils be used to support the theory of organic evolution? _____

_____

_____

How would you determine from which epoch of the Cenozoic a particular group of fossils came? _____

_____

_____

Can you think of any examples, besides those we have given, of the economic utility of fossils? _____

_____

_____

Check the buildings and monuments in the area where you live. What type of rock or fossiliferous rock was used in some of them? _____

_____

_____

WHERE TO COLLECT FOSSILS

Fossils can be collected almost anywhere there are sedimentary rocks. In addition, fossils occur loose in many gravel pits and streambeds and along some beaches. In this section, we briefly describe a few of the more common collecting localities.

Quarries: A quarry is an open pit from which rock is excavated for use in construction, in road work, and for other purposes. Quarries are excellent places to collect fossils, and the geological surveys of many states will provide lists of both currently operating and abandoned quarries. Furthermore, most large-scale topographic maps also show the locations of the quarries in existence when the maps were prepared.

Most quarry operators will not permit you to collect in their quarries while they are being worked, but they may allow you to collect there on weekends and during other periods when they are not in operation. In any case, you must always check first and get permission before doing any collecting.

Many abandoned quarries are excellent places to collect (Figure 4–14). Because of liability laws, however, many quarry owners will not allow you to collect on

Figure 4–14. Students collecting fossils at a quarry.

their property without your first signing release forms. Therefore, again, it is best to check with the owner before doing any collecting.

After you have obtained permission to collect, only the abundance of the fossils and your own skills will limit your collecting. Generally speaking, the best collecting is usually from weathered material, particularly in limestone quarries. Because weathering softens the rock, fossils often appear to be accentuated in weathered rock.

A careful examination of the rubble and piles will give you an idea of the kinds of fossils available, and also of their quality and abundance. You also will want to examine the quarry floor and walls to determine if fossils are abundant enough to try to remove them directly from the rock. This examination will help you determine which units yielded the fossils that you found in the loose quarry rubble.

Highway and Railroad Cuts: Road cuts are excellent places to collect fossils because they have exposed rock layers right beside the road. Cuts along old roads are especially good because those exposures have undergone weathering and erosion, thus producing both weathered rubble along their bases and newly uncovered exposures. Be extremely careful, however, when collecting along a highway. Collecting on interstate highways is forbidden, but, in many places, older roads parallel to them go through the same rock units and yield the same fossils. Driving around an area and checking the piles of rubble along the bottoms of road cuts will allow you to determine quickly whether or not an area is promising. Many field trip guides published by state geological surveys, universities, museums, rock-hound clubs, and so on, describe prominent road cuts and give detailed information about them and the fossils they contain.

Railroad cuts are also potential fossil-collecting sites, particularly where they go through limestones or shales. Many can be reached only by walking and so are not as convenient as road cuts. Because they are more difficult to reach, however, they are less likely to have been heavily collected and may be more productive than road cuts. In addition, most railroad cuts are rather old, so that weathering has had a chance to expose more of their fossils. A word of caution, though. When collecting along railroad cuts, be sure to listen for trains that may still use the rails. It is very easy to become completely absorbed in your collecting and not hear the warning whistle of an oncoming train.

Just as for road cuts, many geologic guidebooks are helpful in locating productive fossil-collecting sites along railroad lines.

Coal Mines and Dumps: Coal mines and dumps are abundant in the Midwest and elsewhere, and provide excellent opportunities to collect both plant and animal fossils. Collecting is especially good in strip mines where vast areas are mined for their easily accessible coal seams. Just as with quarries, permission must be obtained from the mine operator before proceeding to collect.

Some coal layers contain impressions of fossil plants and concretions that, when broken open, may yield plants, insects, or other animal impressions. Spatially associated black shales are also good layers for fossil collecting. Many of the fossils found in these layers have been replaced by pyrite.

Associated gray shale layers, which are the most common rock units in many of these areas, commonly contain excellently preserved plant impressions along their bedding planes (Figure 4–8). These shale layers must be collected fairly soon after being exposed, because a few rains will turn them into sticky piles of clay. Concretions, many of which contain fossils, occur within these shales. The concretions, which must be broken open to see their contained fossils, are best collected from the waste piles. Just as in quarries, the dumps (waste piles) of coal mines often provide the best collecting. It is there that large amounts of material are concentrated and weathered.

Gravel Pits: Gravel pits are places where sand and gravel have accumulated. Many of the deposits in the northern part of the United States were formed as a result of glacial deposition; others represent old river or lake deposits. Gravel pits are found all over the United States and, like quarries and mines, many are marked on topographic maps. Their offices can often be found in the telephone book under "Sand and Gravel."

Permission should be obtained from the pit operator before you begin collecting. Most fossils found in gravel pits will be in the gravel itself and will not occur as loose specimens. Many gravel pits have also yielded bones and teeth of Ice Age animals. Unfortunately for the paleontologist, it is often impossible to tell where the bones and teeth originated.

Natural Exposures: Natural exposures include river banks, cliff faces, beaches, or any other places where nature, rather than humans, has done the excavating. Specimens can be collected from river banks and cliff faces just as they can from highway cuts or quarry faces. There will generally be a small pile of debris at the bottom of the river bank or cliff face, and that is where one should look first to see what types of fossils may be found.

Beaches may yield fossils either from exposed cliffs behind them, or in material that has been transported to them by the water. Take care to establish their source—especially as to whether it is obviously nearby or is unknown and possibly remote.

Other Areas: Just about any exposure, natural or man-made, should be examined for fossils.

Natural exposures, in addition to the ones already mentioned, include caves, sinkholes, peat bogs, and any other places where sedimentary rocks have been exposed. Most man-made exposures are associated with construction, either in progress or recent; for example, building excavations, tunnels, pipelines, trenches, sewer lines, and canals. All are potential sites for fossils.

The best words of advice we can offer are: be alert; don't hesitate to ask around about collecting sites; and good luck!

REVIEW QUESTIONS

Where can fossils be collected? _____

_____

_____

Before collecting in a quarry or mine, what is the first thing you should do?

_____

Where is the best place to look for fossils in a quarry? _____

What is the difference between a quarry, a mine, and a gravel pit? _____

_____

_____

Why are road cuts and railroad cuts good places to collect fossils? _____

_____

_____

_____

What precautions should be taken when collecting? _____

_____

_____

_____

Where have you done most of your fossil collecting? Which places were best? Now that you have finished this chapter and know more about fossils, do you think you will be a better collector? If so, why? _____

_____

_____

_____

_____

_____

_____

REFERENCES AND SUGGESTED ADDITIONAL READING

Arnold, C.A., *Introduction to Paleobotany*. New York: McGraw-Hill, 1947.

Beerbower, J. R., *Search for the Past*. Englewood Cliffs, N.J.: Prentice-Hall, 1960.

Casanova, R., *An Illustrated Guide to Fossil Collecting*. Healdsburg, Calif.: Naturegraph, 1970.

Colbert, E.H., *Dinosaurs*. New York: Dutton, 1961.

Fenton, C.L., and Fenton, M.A., *The Fossil Book*. New York: Doubleday, 1958.

Jones, D.J., *Introduction to Microfossils*. New York: Harper, 1956.

Knowlton, F.H., *Plants of the Past*. Princeton, N.J.: Princeton University Press, 1957.

Kummel, B., and Raup, D., *Handbook of Paleontological Techniques*. San Francisco: W.H. Freeman, 1955.

MacFall, R.P., and Wallin, J., *Fossils for Amateurs*. New York: Van Nostrand, 1972.

Matthews, W.H., III., *Fossils: An Introduction to Prehistoric Life*. New York: Barnes and Noble, 1962.

Moody, R. *A Natural History of Dinosaurs*. Hamlyn, London. 1977.

Moore, R.C. (Editor), *Treatise of Invertebrate Paleontology*. Lawrence, Kansas: University of Kansas Press, and Geological Society of America.

Moore, R.C., Lalicker, C.G., and Fisher, A.G., *Invertebrate Fossils*. New York: McGraw-Hill, 1953.

Murray, M., *Hunting for Fossils*. New York: Macmillan, 1967.

Ransom, J.E., *Fossils in America*. New York: Harper, 1964.

Romer, A., *Man and the Vertebrates*. Baltimore: Penguin, 1954.

Romer, A., *Vertebrate Paleontology*. Chicago: University of Chicago Press, 1966.

Shimer, H.W., and Shrock, R.R., *Index Fossils of North America*. New York: Wiley, 1944.

APPENDIX I

Topographic and Geologic Maps

Maps can be extremely useful to the amateur collector, both for locating collecting sites and for determining the type and age of the rocks at the sites.

Most people are familiar with road maps or atlases available from service stations, automobile clubs, and bookstores. These show the roads of an area and some major natural and cultural features. Although such maps are useful for determining the best way to get to a collecting site, they are not detailed enough to show exactly where the site may be, what the landscape is like, and what the features of the geology are. Consequently, unless you go only on organized field trips, you should obtain and know how to use topographic and geologic maps.

A *topographic map* expresses the third dimension—that is, the relief of the landscape—through the use of lines called *contour lines*. Topographic maps give the elevation of any point on the map and allows you to determine the steepness of slopes, how far different points are from each other, the location of many natural and man-made features, and, in some cases, even individual homesteads.

Topographic maps are available from the United States Geological Survey. For areas east of the Mississippi River, they may be ordered from the Washington Distribution Section, U.S. Geological Survey, 1200 South Eads Street, Arlington, Va. 22202; for areas west of the Mississippi River, orders should be directed to the Denver Distribution Section, U.S. Geological Survey, Federal Center, Denver, Col. 80225. Indexes of available maps, as well as a brochure explaining the symbols used on topographic maps, can be obtained free of charge from these centers. There are also dealers in many cities who stock these maps. The current price (1982) for the most commonly sold topographic maps is $2.00 per map.

Since topographic maps are essentially scaled-down representations of the earth's surface, the first thing you want to do is to determine the scale of the map you are using.

Although maps can be made to almost any scale, most topographic maps are in one of two scales: 1:24,000 (that is, 1 inch on the map equals 24,000 inches on the earth's surface) and 1:62,500 (that is, 1 inch on the map equals 62,500 inches on the earth's surface). These are referred to as 7½-minute and 15-minute quadrangle maps because each map covers 7½ or 15 minutes of longitude and latitude respectively. The scale is given graphically and proportionately in the center at the bottom of each map. The latitude and longitude are recorded on the corners and sides of the map. All maps are printed with north at the top. The name of the map and the area covered (the 7½-minute or 15-minute quadrangle) are located in the upper-left-hand corner. The names of adjoining maps are given along the sides and at the corners of the map.

Interpreting a topographic map involves both practice and imagination. The third dimension (that is, relief) is expressed by the contour lines. When you use a topographic map, remember that what you are looking at is the appearance of the landscape as you would see it if you were flying overhead. Thus the arrangement and spacing of the contour lines determine whether you are seeing a hill or valley, a steep slope or a gentle one, and so on.

What are contour lines, and how do you use them for visualizing the third dimension? Simply defined, contour lines are lines made by connecting points of equal elevation. One way of visualizing this is to imagine a hill with a series of stakes driven into it at an elevation of 100 feet, another series at 200 feet, another at 300 feet, and so on. If, then, you were to connect each series of stakes with a piece of string, what you would see from the air would be a series of concentric lines some distance apart. In the same way, a hill on a topographic map is represented by a series of concentric lines, each line 100 feet higher or lower than the adjacent one. If the slope on one side of the hill were steep and on the other side, gentle, the lines on the steep side of the hill would be close together, whereas those on the gentle side would be more widely spaced. This is true because on a steep slope you have to walk only a short distance to go up or down a given vertical distance, but on a gentle slope you have to walk a much greater distance to cover the same vertical distance.

Another way to visualize contour lines is to actually see their natural analog, by observing the shore zone of a lake or pond whose water level has dropped. Each time the water level drops, a new shoreline is created. After a time, there is a whole series of old shorelines representing the former lake levels. These lines are natural contour lines.

Generally, every fifth contour line on a map is darkened and has its elevation printed on it. The contour interval of the map—that is, the difference in elevation between successive lines—is listed below the map scale. A contour interval of ten feet means each contour line represents an increase or decrease in elevation of ten feet over its adjacent contour. The elevation of certain prominent objects and benchmarks is given in black. A benchmark (*BM* on a map) is an actual marker placed in the ground whose elevation has been determined precisely. Benchmarks are metal plates on a concrete base or in rock. They should never be disturbed.

Figure I–1 shows a drawing of an area together with a topographic map of that area. You can see that where the slopes are steep, the contour lines are close together, and where the slope is gentle, they are widely spaced. Also, stream valleys are marked by a V-shaped or U-shaped series of lines that cross the valleys. The sharp point of the V always points upstream.

With a little practice, you will have no trouble reading and interpreting topographic maps. Next, learn how to locate places, such as collecting sites, on a topographic map.

One way is to use *longitude* and *latitude*. Quadrangle maps (7½- or 15-minute quadrangles) are based on meridians of longitude and parallels of latitude. Meridians of longitude run north to south and divide the earth into wedge-shaped pieces measured east and west of the prime meridian at Greenwich, England. Parallels of latitude run east to west and divide the earth into pieces measured north or south of the equator. Each degree (°) of longitude and latitude can be divided into 60 minutes ('), and each minute can be divided into 60 seconds ("). Therefore, by using a scale, one can locate objects or places on a map precisely merely by designating the location according to its longitude and latitude, such as 115° 12' 42.5" W, 38° 58' 06" S.

Another method for locating points on the earth's surface that is used in the United States (except for the eastern founding states and Texas) is to divide an area into six-mile squares called *townships* (Figure I–2). The townships may be subdivided further into one-mile squares called *sections,* each of which contains 640 acres. Each township contains 36 sections; each section can be divided into *quarter sections;* and each quarter section can be divided further into quarter sections. Actually this division procedure can be repeated as many times as desired, but it generally doesn't go beyond three divisions.

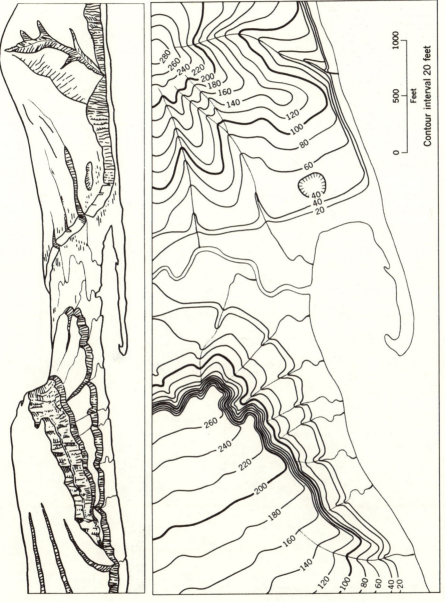

Figure I–1:   (A) Perspective sketch of a landscape. (B) Contour map of A. (Reprinted by permission from R. F. Flint and B. J. Skinner, *Physical Geology.* © 1974 John Wiley & Sons, Inc., New York.)

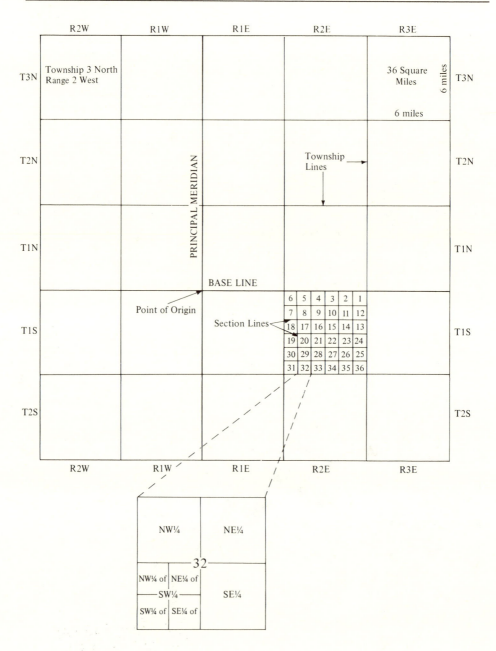

Figure I–2:   U.S. Public Land Survey grid.

Townships are numbered north and south from a baseline, starting with 1. The symbol *T* is used to represent a township on a map. Townships are also numbered east and west from a principal meridian; these are called *ranges,* symbol *R* on a map. Thus, as an example, a township might be Township 4 North and Range 2 East (T4N, R2E on a map), which would be a six-mile square 24 to 30 miles north of the baseline and 12 to 18 miles east of the principal meridian. In turn, a township can be divided into 36 sections, each one mile square. Section 1 is located in the upper-right-hand corner of the township and numbered as shown in Figure I-2. The section number is located in the center of each section on a map. Each section may then be divided into quarter sections (NW¼, NE¼, SW¼, and SE¼), and each quarter section can be further quartered, and so on (Figure I-2). One always lists the smallest quarter first and ends up with the quarter of the section of the township. For example, the *x* in Figure I–2 can be described as being located in the northeast quarter of the southwest quarter of section 32 of Township 1 South and Range 2 East. This can be described in a shorthand notation as NE¼, SW¼, Sec. 32, T1S, R2E. Virtually any location or object can be located very precisely by using this method. Many collecting sites are described in the geologic literature in this way.

A *geologic map* is a topographic map with the geology of the area superimposed on it. Such maps provide a wealth of information for the person who understands it. A geologic map can tell you, for example, the bedrock formation, its age and rock type, and the geologic structure of the area.

For our purposes it is only necessary that you be able to tell the rock unit and its geologic age. This information is provided on the right side of the map where each formation has a color and/or symbol that corresponds to that used to designate it on the map. Thus, if your collecting site is located where the map is colored green and has the symbol K for the green color, the explanation on the right side of the map would show you that you are collecting from Cretaceous aged rocks. For a more detailed explanation of geologic maps and their symbols, refer to an introductory geology textbook, a geologic field methods book, or the appropriate Data Sheet published by the American Geological Institute.

APPENDIX II

Mineral Determinative Tables

Most of the minerals included on the two tables given in this appendix are relatively common and/or widespread. Descriptive data about several additional minerals are tabulated in other publications (see Dietrich, R. V., *Mineral Tables: Hand-Specimen Properties* . . . New York: McGraw-Hill, 1969; Larson, E. S., and Berman, H., *The Microscopic Determination of the Nonopaque Minerals,* 2nd ed. U.S. Geological Survey Bulletin 848, 1934; and Short, M. N., *Microscopic Determination of the Ore Minerals*, U.S. Geological Survey Bulletin 825, 1931.)

Tables II–1 and II–2 include minerals with metallic lusters and with nonmetallic lusters, respectively. Minerals with submetallic lusters are included in both tables, so you will find them even if you have considered them as definitely metallic or nonmetallic in appearance. In each of the two tables, the minerals are arranged in order of least to greatest hardness; within groups of minerals having essentially equal hardnesses, the minerals are listed in order of least to greatest specific gravity. The relatively common colors for each mineral are given on the left side of the table. Chemical formulas, crystal systems, and remarks relating to other characteristics and/or occurrences are also given.

The chemical formulas are given because we consider it wise for you to think about minerals as chemical compounds from the onset. To aid you in this pursuit, we have included in the tables' *Remarks* columns such designations as "Pb-gray," meaning lead-gray; "Cu-deps," meaning copper deposits; and "Mn-oxides," meaning manganese oxides. If you need it, Appendix VI provides additional information about chemical symbols and formulas.

The crystal systems are indicated by abbreviations as follows: isometric—Isom, tetragonal—Tetr, hexagonal—Hexa, orthorhombic—Orth, monoclinic—Mncl, and triclinic—Trcl. Specific gravity values are given as single figures or as ranges. In the *Remarks* column, many abbreviations are used. Those whose meanings may not be obvious are as follows:

Access   accessory mineral
Admn   adamantine
Aggs   aggregates
Alk(s)   alkalic (e.g., rock) *or* alkaline (e.g., taste)

Alter   alteration product
An   anorthite content (e.g., An30 indicates plagioclase with the composition anorthite—30 percent and albite—70 percent)

Anal   analysis
Aq-reg   aqua regia
Assoc   associated (with)
Attkd   attacked (by)
Bldd   bladed
Blds   blades
Blk   black
Blu   blue
Brtl   brittle
Brwn   brown
Calc   calcareous
Clvg(s)   cleavage
Cntct   contact (as applied to rocks metamorphosed near contacts with igneous rocks)
Conc   concentrated
Conch   conchoidal
Dcmp   decomposes in
Dep(s)   deposit
Dissem   disseminated
Dk   dark
Dodec   dodecahedral
Dolo   dolomite (the rock)
Duct   ductile
Eff   effervescent
Efflor   effluorescent (ce)
Elast   elastic
Expos   exposure *or* exposed
Fibr   fibrous
Fl   flame
Flex   flexible (e.g., to flexible sheets)
Flks   flakes
Fluo   fluoresces (in ultraviolet light, unless otherwise indicated) *or* fluorescent
Fm(s)   form *or* formation
. . . fm   form (as a suffix; e.g., collofm)
Fol   foliated

Fs   ferrosilite content (e.g., Fs30 indicates pyroxene of the enstatite-ferrosilite series with a composition of ferrosilite—30 percent and enstatite—70 percent)
Gelat   gelatinizes (in HCl, unless otherwise noted)
Glbl   globular
Gldn   golden
Grn(s)   grain
Grnlr   granular
Grp   group
Grsy   greasy
Hex   hexagonal
Hi   high
Hi-t   high-temperature (e.g., as referred to metamorphic processes)
Ig   igneous (rocks)
Incrust(s)   incrustation
Inelast   inelastic
Irid   iridescent
Irreg   irregular
. . . lk   like (as a suffix; e.g., lenslk)
Lo   low
Lo-t   low-temperature (e.g., as referred to metamorphic processes)
Ls   limestone
Lw   long-wave (as referred to ultraviolet radiation)
Mag   magnetic
Mal   malleable
Meso-t   mesothermal
Met   metallic
Meta   metamorphism *or* metamorphic
Min(s)   mineral(s)
Mod-t   moderate-temperature (e.g., as referred to metamorphic processes)
Msv   massive

Octah   octahedral *or* octahedra
Opq   opaque
Or   orange
Peg(s)   pegmatite
Perf   perfect
Phospho   phosphorescent
Plts   plates
Plty   platy
Prism   prismatic
Prod   product
Qtz   quartz
Rdactv   radioactive
Resin   resinous
Rhomb   rhombohedral
Rk(s)   rock
S-vns   sulfide (not sulfur) veins
Scndry   secondary
Sect   sectile
Sed   sediment, sedimentary, *or* sedimentary rock
S.G.   specific gravity
Shpd   shaped
Sil   siliceous
Sol   soluble (in)

Stalac   stalactitic
Stl   steel
Strk   streak
Sufs   sulfides
Sw   short-wave (ultraviolet radiation)
Tarn   tarnish *or* tarnished
Tblr   tabular
Tetra   tetrahedral *or* tetrahedra
Thermolum   thermoluminescent
Tribolum   triboluminescent
Unevn   uneven
Vitr   vitreous
Vn(s)   vein
Vol(s)   volcanic *or* volcanic rock
Wdsprd   widespread
Wht   white
Wxed   weathered
Wxing   weathering
Xl(s)   crystal
Yel   yellow
Zeol-assoc   typical zeolitic association

In the *Remarks* column, commas are used to separate terms of the same category (for example, Vitreous, adamantine) whereas periods are used between categories (for example, Vitreous. Fluorescent). Dashes may be thought of as different prepositions; (for example, Vitr-admn means the luster ranges from vitreous *to* adamantine, while Sol-HCl means soluble *in* hydrochloric acid. The following symbols have their standard meanings: $\sim$ for approximately; $>$ for greater than; and $<$ for less than.

HOW TO USE THESE TABLES

A suggested procedure for running down a mineral by using these tables is the following:

1. Note whether the mineral has a metallic or nonmetallic luster.
2. Check the color.
3. Determine the Mohs hardness.

After these steps, go to the appropriate part of the appropriate table. In nearly all cases, you will have narrowed the possibilities to only a few minerals. Observe or determine any information that would appear to be diagnostic on the basis of the Remarks.

Here is an example of this procedure:

1. The mineral is nonmetallic; therefore, use Table II–2.
2. It is white.
3. It has a hardness of about 3.

According to the table, each of the relatively common white minerals with hardnesses between 2 and 4 may be identified readily because of such properties as its characteristic taste, its reaction with dilute HCl, or its fluorescence. Try, for example, a white or colorless piece of calcite or a grain of halite (common table salt), both of which would fall in this group, to see if you can run them down with this procedure.

Try several minerals! The more you identify, the more adept you will become. Sooner or later you will probably start to recognize some minerals upon sight or by just making one or two checks that you remember. Starting this way, many amateur mineral collectors have built up their expertise and can now readily and rapidly identify several scores of minerals.

Always remember, however, that the mineral you are trying to identify may not be included on these tables. If you find that your mineral just does not fit, it will be wise to have an advanced amateur or a professional mineralogist or petrologist check it for you.

Table II–1 Minerals with Metallic Luster

Hardness	Name, Formula	Crystal System	Specific Gravity	Remarks	Gray/Black	Red/Orange	Yellow/Green	Blue/Purple
1–1½	Molybdenite MoS_2	Hexa	4.62–4.73	Opq. Pb-gray. Strk-green (glazed porcelain), blue-black (paper). Tblr, fol msv, scales. Demp-HNO₃. Access in granites, pegs, vns, cntct meta rks.	X			X
1–2	Graphite C	Hexa	2.09–2.23	Opq. Scales Strk-black. 1 perf clvg to inelastic plts. Meta of carbonaceous rks.	X			
1½–2	Orpiment As_2S_3	Mncl	3.49	Fol, fibr, grnlr. Resin. 1 clvg-inelastic. Lo-t vns, alter prod.		X	X	
1½–2	Covellite CuS	Hexa	4.60–4.76	Some irid. Gray strk. 1 perf clvg-flex plts. Alter of Cu-sulfides.	X		X	X
2	Stibnite Sb_2S_3	Orth	4.61–4.65	Opq. Pb-gray, blk-irid tarn. Striated bent blds. Sol-HCl. Lo-t vns.	X			
2–2½	Cinnabar HgS	Hexa	8.09	Scarlet strk. Admn-dull. Near vols + hot springs-vns + incrusts.	X	X		
2–6½	Pyrolusite MnO_2	Tetr	4.40–5.08	Opq. Fibr, grnlr. Prism clvg. Bog and residual Mn-deps-common.	X			X
2½–3	Chalcocite Cu_2S	Orth	5.50–5.80	Opq. Msv. Conch. Sect. Sol-HNO₃. Enriched zones atop sulfide deps.	X			

				H	Mineral / Formula	Sys.	G	Remarks
X				$2\frac{1}{2}$–3	Galena PbS	Isom	7.57–7.59	Opq. Pb-gray. Cubic, msv. Perf cubic clvg. Dcmp-HNO_3. Sulfide vns.
	X			$2\frac{1}{2}$–3	Copper Cu	Isom	8.94	Opq. Branching xls. Mal, duct. Sol-HNO_3. Basaltic vols. Veins.
X		X		$2\frac{1}{2}$–3	Silver Ag	Isom	10.50	Opq. Gray to blk tarn. Elongate fms. Mal, duct. Sol-HNO_3. Veins
	X			$2\frac{1}{2}$–3	Gold Au	Isom	19.31	Opq. Branching xls, nuggets. Mal, duct. Veins and placers.
X				3	Enargite Cu_3AsS_4	Orth	4.40–4.50	Opq. Tblr, msv, grnlr. Prism zone-striated. 3 good, 1 poor clvgs. Sol-aq-reg. Mod-t to Lo-t vein + replacement sulfide deps.
X	X			3	Bornite Cu_5FeS_4	Isom	5.06–5.08	Opq. Purplish irid tarn. Gray-blk strk. Msv. Sol-HNO_3. Cu-deps.
	X			3–$3\frac{1}{2}$	Millerite NiS	Hexa	5.30–5.70	Opq. Brass-bronze yel, grn-blk strk. Hairlk xls. Lo-t, in cavities.
X	X			3–$4\frac{1}{4}$	Tennantite $(Cu,Fe)_{12}As_4S_{13}$	Isom	4.62	Softer and lighter than tetrahedrite. Tetra, msv. Dcmp-HNO_3. Vns.
X	X			3–$4\frac{1}{4}$	Tetrahedrite $(Cu,Fe)_{12}Sb_4S_{13}$	Isom	4.99	Harder and heavier than tennantite. Tetra, msv. Demp-HNO_3. Wdsprd-vns.
X	X	X		$3\frac{1}{2}$–4	Sphalerite $(Zn,Fe)S$	Isom	3.90–4.10	Resin. Curved xl faces common. Fluo-or. Tribolum. Sol-HCl. Vns.
	X			$3\frac{1}{2}$–4	Chalcopyrite $CuFeS_2$	Tetr	4.10–4.30	Opq. Brassy, irid tarn, grn-blk strk. Sol-HNO_3. Meso-to Hi-t vns.
X	X			$3\frac{1}{2}$–4	Cuprite Cu_2O	Isom	6.14	Submet(some). Octah, hairlk, msv. Common oxidized zones of Cu-deps.

Table II–1 Minerals with Metallic Luster

Hardness	Name, Formula	Crystal System	Specific Gravity	Remarks	Gray/Black	Red/Orange	Yellow/Green	Blue/Purple
3½–4½	Pyrrhotite $Fe_{1-x}S$	Hexa	4.58–4.79	Opq. Bronzy, tarn. Msv. Unevn. Mag. Dcmp-HCl. Basalts, S-veins.		X	X	X
4	Manganite $MnO(OH)$	Mncl	4.32–4.34	Submet. Red-brwn strk. Striated prisms. 3 clvgs. Sol-HCl. Lo-t vns.	X			
4	Huebnerite $MnWO_4$	Mncl	7.12	Submet. Tarn-irid. Prism-striated, parallel to radiating xl grps. 1 clvg. Dcmp-aq-reg, H_2SO_4 or HCl (slowly). Diverse-vns, cntct meta, placers.		X	X	
4–4½	Wolframite $(Fe,Mn)WO_4$	Mncl	7.31	Submet. Tarn-irid. Prism-equant—striated, msv-grnlr, etc. 1 clvg. Dcmp-aq-reg, H_2SO_4 or HCl (slowly). Diverse-vns, cntct meta, placers.	X			X
4–4½	Platinum Pt	Isom	19.00	Opq. Grains and scales. Mal, duct. Ultrabasic ig rks and placers.	X			
5–5½	Goethite alpha-$FeO(OH)$	Orth	3.30–4.29	Submet. Yellowish strk. Msv, fibr, etc. Silky-admn. Sol-HCl. Wxing product.	X	X	X	
5–5½	Nickeline NiAs	Hexa	7.78	Opq. Pale Cu-red, gray tarn, brwnish strk. Sol-aq-reg. As, S-vns.	X	X		
5–6	Bronzite $(Mg,Fe)_2Si_2O_6$	Orth	3.30–3.45	Submet. (Bronzy). Fs12-30. In basic and ultrabasic ig rks, some metas.		X	X	

	H	Mineral	G	System	Remarks
X	5–6	Ilmenite FeTiO$_3$	4.68–4.76	Hexa	Submet. Opq. Tblr. Vns + dissem deps assoc with gabbros + diorites.
X X	5–6	Hematite alpha-Fe$_2$O$_3$	5.26	Hexa	Met-earthy. Strk-red. Plty, grnlr. Sol-HCl. Sed Fe-fms, rare in vns.
X X	5–6	Uraninite UO$_2$	10.63	Isom	Submet. Opq. Rarely octah; colloform ("pitch blende"), etc., msv. Grsy-dull. Rdactv. Pegs, hi and mod-t vns.
X	5½	Chromite FeCr$_2$O$_4$	4.50–4.80	Isom	Submet. Msv, grnlr. Mag (feebly). With olivine-rich (commonly serpentinized) ig rks.
X X ... X	5½	Cobalite CoAsS	6.33	Isom	Opq. Typically reddish Ag-wht, also violet stl gray + dk gray. Habit-like pyrite. Cubic clvg. Dcmp-HNO$_3$. Hi-t dissem deps + vns.
X X X X	5½–6	Anatase TiO$_2$	3.90	Tetr	Submet. Pale yel strk. Acute pyramidal. Admn. Basal + pyramidal clvg. Converts to rutile on heating. Vns, access-ig rks, detrital.
X ... X	5½–6	Arsenopyrite FeAsS	6.55–6.85	Mncl	Opq. Ag-wht, stl-gray. Prism. Dcmp-HNO$_3$. Wdsprd-e.g., ore deps, pegs.
X	5½–6½	Magnetite FeFe$_2$O$_4$	5.18	Isom	Submet. Blk strk. Octah, grnlr. Octah parting. Difficultly sol-HCl. Strongly magnetic. Wdsprd.
X X	6	Columbite (Fe,Mn)Nb$_2$O$_4$	5.15–5.25	Orth	Submet. In series with tantalite. Red-blk strk. Tarn-irid. Tblr, prism. 2 clvgs. Brittle. Granitic pegs.

Table II–1 Minerals with Metallic Luster

Hardness	Name, Formula	Crystal System	Specific Gravity	Remarks	Gray/Black	Red/Orange	Yellow/Green	Blue/Purple
6–6½	Rutile TiO_2	Tetr	4.21–4.25	Submet. Prism-striated. Admn. Poor clvgs. Vns, access-meta and ig rks.	X	X	X	X
6–6½	Marcasite FeS_2	Orth	4.89	Opq. Sn-wht. Bronze-yel. Tblr, cockscomb-lk. Lo-t, acid deposition.			X	
6–6½	Pyrite FeS_2	Isom	4.82–5.02	Opq. Brassy. Cubes, dodec, msv. Unevn. Sol-(powder)-HNO_3. Wdsprd.			X	
6–6½	Tantalite $(Fe,Mn)Ta_2O_6$	Orth	7.90–8.00	Submet. In series with columbite, which see. Pegs.	X	X		
6–7	Cassiterite SnO_2	Tetr	6.99	Submet. Radial concretionary masses. Admn-dull. Hi-t vns, greisens.	X	X	X	

Table II-2 Minerals with Nonmetallic Lusters

Colorless/White	Gray/Black	Red/Orange	Yellow/Brown	Green/Blue/Purple	Hardness	Name, Formula	Crystal System	Specific Gravity	Remarks
X	X			X	1	Talc $Mg_3Si_4O_{10}(OH)_2$	Mncl	2.58–2.83	Fol msv. Basal clvg. Sectile. Soapy feel. Pearly. Schist, meta ig.
X			X	X	1–2	Vermiculite $(Mg,Fe,Al)_3(Al,Si)_4O_{10}(OH)_2 \cdot 4H_2O$	Mncl	2.20–2.40	Bronzy. Micaceous. Exfoliates with swelling on heating. Attkd-acid (silica residue). Alter prod of biotite (etc.)-hydrothermal or wxing.
X	X		X	X	1–2	Montmorillonite $(Na,Ca)_{0.33}(Al,Mg)_2Si_4O_{10}(OH)_2 \cdot nH_2O$	Mncl	2.20–2.70	Irreg and hex-shaped plts. "Swelling clay." Alter-hi Mg and Ca, lo-K rks (alkaline conditions). Soils, bentonites, etc.
X			X	X	1–2	Pyrophyllite $Al_2Si_4O_{10}(OH)_2$	Mncl	2.65–2.90	Lamellar. Clvg-inelast plts. Partly dcmp-H_2SO_4. Vns, few schists.
X	X		X	X	1–3	Clay Complex hydrous aluminum silicates	Mncl	1.85–3.00	Name given several hydrous Al silicates with alkalies or alkaline earths ± Mg or Fe. Names unsettled. Most distinguished by nonmegascopic means. Fine grained aggs (compact, mealy, etc.). Earthy-pearly. Unctuous. Plastic when wet. Dehydrated on heating. Some are absorbent. Wxing prods, hydrothermal alter, detrital, diagenic.

Table II–2 Minerals with Nonmetallic Lusters

	Hardness	Name, Formula	Colorless/White	Gray/Black	Red/Orange	Yellow/Brown	Green/Blue/Purple	Crystal System	Specific Gravity	Remarks
	$1\frac{1}{2}$	Ice H_2O	X				X	Hexa	0.92	Liquid above 0°C (32°F) and below 100°C (212°F).
	$1\frac{1}{2}$–2	Nitratite $NaNO_3$	X	X	X	X		Hexa	2.24–2.29	Incrust, rhomb. Vitr. Rhomb + other clvgs. Cooling taste. In arid soils in protected locations.
	$1\frac{1}{2}$–2	Vivianite $Fe_3(PO_4)_2 \cdot 8H_2O$	X				X	Mncl	2.67–2.69	Prism, rounded, glbl. Vitr-dull. Clvgs. Fibr-flex. Darkens in exposure. Sol-acids. Scndry of diverse occurrences.
	$1\frac{1}{2}$–2	Realgar AsS			X	X		Mncl	3.48–3.56	Red, or yel. Resin-grsy. Dcmp-HNO_3. With As-mins in Pb, Ag, Au-vns.
	$1\frac{1}{2}$–2	Covellite CuS		X	X	X	X	Hexa	4.60–4.76	Some irid. Gray strk. 1 perf clvg-flex plts. Alter of Cu-sulfides.
	$1\frac{1}{2}$–$2\frac{1}{2}$	Sulfur S		X	X	X		Orth	2.07	Conch-unevn. Brtl. Burns-blu fl + acrid odor. Sol-CS_2. Vols and sed.
	2	Melanterite $FeSO_4 \cdot 7H_2O$	X	X		X	X	Mncl	1.84–1.90	Stalac, crusts, etc. Vitr. Clvgs. Sweet, astringent, metallic taste. Expos (dry air)-yel-wht + opq. Efflorescence by oxidation of sulfs.
	2	Sylvite KCl	X	X	X	X	X	Isom	1.98–2.00	Cubic, grnlr. Vitr. Cubic clvg. Bitter taste. Sed evaporite deps.

Tests	H	Mineral / Formula	System	G	Remarks
X X	2	Niter KNO_3	Orth	2.10–2.11	Crusts, grnlr. Vitr. Clvgs. Salty, cooling taste. Efflor-caves, etc.
X X X	2	Halite $NaCl$	Isom	2.17	Cubic, grnlr. Vitr. Cubic clvg. Salty taste. Sed evaporite deps.
X X X	2	Gypsum $CaSO_4 \cdot 2H_2O$	Mncl	2.31–2.32	Tblr, msv. Subvitr-pearly. 3 clvg, 1 perf (flex, inelast). Sol-HCl. Sedimentary bedded deps, fumarolic, efflorescences, etc.
X X	2	Glauconite $(K,Na)(Fe,Al,Mg)_2-(Si,Al)_4O_{10}(OH)_2$	Mncl	2.40–2.95	Plts in round grains. Attkd-HCl. Marine deps—greensands.
X	$2-2\frac{1}{2}$	Epsomite $MgSO_4 \cdot 7H_2O$	Orth	1.68	Crusts. Vitr-earthy. Clvgs. Bitter, salty. Metallic. Common efflor.
X X	$2-2\frac{1}{2}$	Borax $Na_2B_4O_7 \cdot 10H_2O$	Mncl	1.71–1.72	Habit-lk pyroxene. Vitr-earthy. Clvgs. Sweetish. Sol-H_2O (to alk solution). Borax deps (saline lake deps, etc.).
X X X X	$2-2\frac{1}{2}$	Kaolinite $Al_2Si_2O_5(OH)_4$	Trcl	2.60–2.68	Alter (acid conditions) of feldspars, etc. Hydrothermal + wxing.
X X X	$2-2\frac{1}{2}$	Phlogopite $KMg_3Si_3AlO_{10}(F,OH)_2$	Mncl	2.76–2.90	Mg-mica. Attkd-HCl. Cntct meta-ls. Ultrabasic ig rks.
X X	$2-2\frac{1}{2}$	Autunite $Ca(UO_2)_2(PO_4)_2 \cdot 10\text{-}12H_2O$	Tetr	3.10–3.20	Tblr, subparallel, fol, scaly aggs, serrate crusts. Vitr, pearly. 2 clvgs. Sol-acids. Fluo-yel-green. Alter prod of U mins (pegs, vns).
X X X	$2-2\frac{1}{2}$	Cinnabar HgS	Hexa	8.09	Scarlet strk. Admn-dull. Near vols and hot springs-vns and incrusts.

Table II-2 Minerals with Nonmetallic Lusters

Hardness	Name, Formula	Crystal System	Specific Gravity	Remarks	Colorless/White	Gray/Black	Red/Orange	Yellow/Brown	Green/Blue/Purple
2–3	Chlorite Mg-Fe hydrous aluminum silicates	Mncl	2.60–3.30	Group name. Member names unsettled. Only few rare varieties megascopically distinguishable. Dissem flks. Pearly. 1 perf clvg to flex, inelast folia. Dcmp-in boiling H_2SO_4. Scndry-hydrothermal alter, lo grad meta, detrital, authigenic.	X	X		X	X
2–3	Carnotite $K_2(UO_2)_2(VO_4)_2 \cdot 3H_2O$	Mncl	4.10	Powder, crusts, micro xls. Dull-silky. 1 clvg. Sol-acids. Scndry-e.g., sandstone matrices.				X	X
2–4	Mica Complex hydrous aluminum silicates	Mncl	2.40–3.40	Group name. Some species distinguishable on basis of color. Series are not complete. Hex-shaped xls, dissem. Perf clvg to flex, elast flks. Percussion + pressure figures may be imposed on flks. Chief occurrences are noted under species: biotite, muscovite, phlogopite, lepidolite, and paragonite.	X	X		X	X
2–4	Bauxite		2.00–3.50	Field term for materials rich in hydrous aluminum oxides.	X	X		X	X
2–6½	Wad		2.80–4.40	Field term for hydrous Mn-oxides of unknown identity (Mn-deps.)		X		X	

Marks	H	Name / Formula	System	G	Remarks
X X	$2\frac{1}{2}$	Chrysotile $Mg_3Si_2O_5(OH)_4$	Mncl	2.20	Fibrous, some is asbestiform. Scndry, commonly in vns.
X	$2\frac{1}{2}$	Paragonite $NaAl_2(Si_3Al)O_{10}(OH)_2$	Mncl	2.85	Na-mica. Distinguished from muscovite by chem or X-ray anal. Wdsprd in schists and phyllites, also in vns.
X X	$2\frac{1}{2}$–3	Biotite $K(Mg,Fe)_3(Al,Fe)Si_3O_{10}(OH,F)_2$	Mncl	2.70–3.30	Mg-, Fe-mica. Bleached and attkd-H_2SO_4. Common-e.g., acid ig, pegs, meta.
X	$2\frac{1}{2}$–3	Sericite	Mncl	2.77–2.88	Fine-grained white (Na or K) mica. Deuteric-ig rks, meta rks.
X	$2\frac{1}{2}$–3	Crocoite $PbCrO_4$	Mncl	5.96–6.02	Prism, msv, etc. Admn-vitr. Clvgs. Conch. Sectile. Scndry-in gossans.
X	$2\frac{1}{2}$–3	Vanadinite $Pb_5(VO_4)_3Cl$	Hexa	6.88	Prism, hairlk. Resin. Conch. Some show concentric zoning. Sol-HNO_3 (yel), HCl (green). Oxidized zones of Pb deps.
X X	$2\frac{1}{2}$–3	Wulfenite $PbMoO_4$	Tetr	6.50–7.00	Tblr, dipyramidal, msv, grnlr. Resin-admn. Clvgs. Unevn. Dcmp-HCl, HNO_3, Sol-H_2SO_4, alks. Scndry-oxidized zones of Pb- and Mo-bearing rks.
X	$2\frac{1}{2}$–$3\frac{1}{2}$	Gibbsite $Al(OH)_3$	Mncl	2.30–2.42	Tblr, mammillary. Pearly-vitr. 1 clvg. Claylk odor when damp. Wxing prod. Bauxite deps, lo-t vns.
X X	$2\frac{1}{2}$–$3\frac{1}{2}$	Serpentine $(Mg,Fe,Al)_3(Si,Al)_4O_{10}(OH)_2 \cdot 4H_2O$	Mncl	2.50–2.60	Group name. Nomenclature for members is unsettled. Msv-asbestifm. Waxy-grsy. Gelat. Some fluo-cream yel. Alter prod after Mg-silicates (e.g., olivine), vns.
X	$2\frac{1}{2}$–$3\frac{1}{2}$	Jarosite $KFe_3(SO_4)_2(OH)_6$	Hexa	2.91–3.26	Crusts, grnlr, etc. Resin. 1 clvg. Sol-HCl. On Fe-bearing ores and rks.

Table II–2 Minerals with Nonmetallic Lusters

Colorless/White	Gray/Black	Red/Orange	Yellow/Brown	Green/Blue/Purple	Hardness	Name, Formula	Crystal System	Specific Gravity	Remarks
X	X			X	$2\frac{1}{2}$–4	Muscovite $KAl_2(Si_3Al)O_{10}(OH)_2$	Mncl	2.77–2.88	K-mica. Common rk-forming min-e.g., acid ig, pegs, greisens, cntct metasomatized (fluorine) zones, regional meta-argillaceous rks.
X		X			$2\frac{1}{2}$–4	Lepidolite $K(Li,Al)_3(Si,Al)_4O_{10}^-(F,OH)_2$	Mncl	2.80–2.90	Li-mica. Tabular xls, aggs of small flks. Distinguished from rose-colored muscovite by chem or X-ray anal. Rare, Li-bearing pegs.
X	X	X	X	X	3	Calcite $CaCO_3$	Hexa	2.71	Various trigonal xl habits, msv. Vitr-pearly. Rhomb clvg. Some is fluo, phospho, thermolum. Eff-dilute HCl. Wdsprd-vns, diverse rks.
X	X			X	3–$3\frac{1}{2}$	Laumontite $CaAl_2Si_4O_{12}\cdot4H_2O$	Mncl	2.20–2.30	Perf clvgs-3 directions in 1 zone. Gelat. Cavities in several rks.
X	X			X	3–$3\frac{1}{2}$	Celestite $SrSO_4$	Orth	3.96–3.98	Typically colorless-bluish. Tblr, etc. Vitr-pearly. 3 clvgs. Some fluo, thermolum. Sol-H_2O (slightly). Sed rks (esp gypsum + ls), vns.
X	X			X	3–$3\frac{1}{2}$	Witherite $BaCO_3$	Orth	4.29	Pseudohex, glbl, etc. Vitr. Fluo (etc) - lk aragonite. Sol-HCl. Vns.
X	X	X	X	X	3–$3\frac{1}{2}$	Barite $BaSO_4$	Orth	4.50	Tblr, msv, etc. Vitr-pearly. Clvgs. Some fluo, phospho, thermolum. Some fetid when rubbed. Common-vns, sed rks, cavities in ig rks.

		Hardness	Mineral / Formula	System	G	Remarks
X X	X X	$3–3\frac{1}{2}$	Greenockite CdS	Hexa	4.90	Hemimorph pyramidal xls. Admn-resin. Sol-HCl. Coatings on Zn-mins.
X X	X X	$3–3\frac{1}{2}$	Cerussite $PbCO_3$	Orth	6.53–6.57	Twinned clusters, msv. Admn-resin. Clvgs. Fluo-yel (X-rays and lw ultraviolet). Oxidized zones of Pb-deps.
X X	X X	3–4	Wavellite $Al_3(PO_4)_2(OH,F)_3 \cdot 5 H_2O$	Orth	2.36	Radial fibr, stellate, crusts. Vitr-pearly. Clvgs. Sol-acids. Scndry-wdsprd-openings in diverse rks, limonite, phosphate rks, etc.
X X X	X X	3–4	Stilpnomelane $K(Fe,Al)_{10}Si_{12}O_{30}(OH)_{12}$	Mncl	2.59–2.96	Micaceous (but also 2nd clvg perpendicular to 1st). Sol-HF + H_2SO_4(1/1). Lo-grd meta rks.
X X X X	X X	$3–5\frac{1}{2}$	Zeolite Na +/or Ca hydrous aluminum silicates	Diverse 1.90–2.45		Group of min families with somewhat related appearances, occurrences, and compositions. Many are indistinguishable macroscopically. Diverse habits (e.g., bundles, fibrous). Pearly, glassy. Subconch. Some fluo-orange or yel-green. Continuous, in part reversible, dehydration (i.e., water can be expelled without destroying xl structure). Scndry in open spaces; commonly two or more occur together in cavities in basaltic rks—this occurrence is referred to herein as the zeolite association (abbreviated "zeol-assoc").
X X X	X X	$3\frac{1}{2}$	Anhydrite $CaSO_4$	Orth	2.98	Msv, grnlr, fibr. Vitr-pearly. 3 clvgs. Sol-acids. Sed rks.
X X X	X X	$3\frac{1}{2}$	Strontianite $SrCO_3$	Orth	3.64–3.78	Spear-lk, msv. Vitr. Clvgs. Fluo lk aragonite. Sol-HCl. Vns.

Table II–2 Minerals with Nonmetallic Lusters

Hardness	Name, Formula	Crystal System	Specific Gravity	Remarks	Colorless/White	Gray/Black	Red/Orange	Yellow/Brown	Green/Blue/Purple
3½–4	Heulandite (Na,Ca)$_{2-3}$Al$_3$(Al,Si)$_2$Si$_{13}$O$_{36}$·12H$_2$O	Mncl	2.10–2.20	Plty grp. Tblr-broad central portions (coffinlike). Zeol-assoc.	X	X		X	
3½–4	Stilbite Sb$_2$S$_3$	Mncl	2.10–2.20	Plty grp. Sheaf-like aggs. 1 good clvg. Dcmp-HCl. Zeol-assoc.	X		X	X	
3½–4	Dolomite CaMg(CO$_3$)$_2$	Hexa	2.84–2.86	Rhomb-curved, grnlr. Vitr-pearly. Rhomb clvg. Some fluo. Eff-warm HCl if powdered. Sed rks, xls in cavities, etc.	X	X	X	X	X
3½–4	Alunite KAl$_3$(SO$_4$)$_2$(OH)$_6$	Hexa	2.60–2.90	Grnlr, msv. Vitr-pearly. Clvgs. Sol-H$_2$SO$_4$. "Alunitized" rks.	X	X	X	X	X
3½–4	Aragonite CaCO$_3$	Orth	2.94–2.95	Acicular, radial fibr, stalactitic. Vitr. Clvgs. Fluo (ultraviolet, X-rays, electron beams), thermolum. Eff-acids. Lo-t deposition.	X	X	X	X	X
3½–4	Malachite Cu$_2$(CO$_3$)(OH)$_2$	Mncl	3.60–4.05	Incrust, mammillary. Vitr-dull. Clvgs. Sol-acids. Oxidized zones-Cu deps.			X		X
3½–4	Rhodochrosite MnCO$_3$	Hexa	3.70	Grnlr, botryoidal. Vitr-pearly. Rhomb clvg. Eff-warm acid. Vns.		X	X		
3½–4	Azurite Cu$_3$(CO$_3$)$_2$(OH)$_2$	Mncl	3.77	Tblr-aggs. Vitr. Clvgs. Sol-acids, etc. Scndry zones of Cu-deps.					X

	H	Mineral		Syst	G	Remarks
X X	$3\frac{1}{2}$–4	Cuprite	Cu_2O	Isom	6.40	Submet (some). Octah, hairlk, msv. Common-oxidized zones of Cu-deps.
X X X X X	$3\frac{1}{2}$–4	Pyromorphite	$Pb_5(PO_4)_3Cl$	Hexa	7.00–7.08	Prism, glbl, etc. Resin. Rhomb clvg. Unevn. Sol-HNO_3. Oxidized zones of Pb deps.
X X X	$3\frac{1}{2}$–4	Mimetite	$Pb_5(AsO_4)_3Cl$	Hexa	7.24	Prism, acicular, glbl, etc. Resin. Rhomb clvg. Unevn. Sol-HNO_3. Oxidized zones of Pb-deps.
X X	$3\frac{1}{2}$–4	Magnesite	$MgCO_3$	Hexa	2.98–3.02	Grnlr, rhomb. Vitr. Rhomb clvg. Fluo-green, blue. Sol-warm HCl. Alteration of Mg-rich rks.
X X X X X	$3\frac{1}{2}$–$4\frac{1}{2}$	Siderite	$FeCO_3$	Hexa	3.95–3.97	Tarn-irid. Msv. Rhomb, etc. Vitr. Rhomb clvg. Sol-hot acid. Sed, vns.
X X X X X	4	Fluorite	CaF_2	Isom	3.18	Cubic, msv. Vitr. Octah clvg. Some fluo-violet. Dcmp-H_2SO_4. Wdsprd-vns, cavities, disseminated in ig rks, etc.
X	4	Manganite	$MnO(OH)$	Mncl	4.32–4.34	Submet. Red-brwn strk. Striated prisms. 3 clvgs. Sol-HCl. Lo-t deps.
X	4–$4\frac{1}{2}$	Variscite	$AlPO_4·2H_2O$	Orth	2.20–2.57	Nodules, crusts—some opaline. Waxy. 2 clvgs. Splintery-conch. Sol-alkalies. Near-surface deposition in cavities (e.g., breccias).
X X X X X	4–$4\frac{1}{2}$	Smithsonite	$ZnCO_3$	Hexa	4.40–4.45	Botryoidal, msv. Vitr-pearly. Rhomb clvg. Some fluo-green, blue. Eff-acids. Scndry-oxidized zones of Zn-deps.
X X	4–$4\frac{1}{2}$	Wolframite	$(Fe,Mn)WO_4$	Mncl	7.31	Submet. Tarn-irid. Prism-equant—striated, msv-grnlr, etc. 1 clvg. Dcmp-aq-reg, H_2SO_4 or HCl (slowly). Diverse-vns, cntct meta, placers.
X X X	4–$5\frac{1}{2}$	Limonite			2.70–4.30	Field term for hydrous iron oxides of unknown identities.

Table II-2 Minerals with Nonmetallic Lusters

Colorless/White	Gray/Black	Red/Orange	Yellow/Brown	Green/Blue/Purple	Hardness	Name, Formula	Crystal System	Specific Gravity	Remarks
X			X	X	$4\frac{1}{2}$	Chabazite $CaAl_2Si_4O_{12}\cdot6H_2O$	Hexa	2.05–2.10	Rhombs (nearly cubes). Rhomb clvg. Dcmp-HCl. Zeol-assoc, rk cavities.
X			X	X	$4\frac{1}{2}$–5	Apophyllite $KCa_4(Si_4O_{10})_2(OH,F)\cdot8H_2O$	Tetr	2.33–2.37	Vitr sq prisms, pearly bases. Basal + poor prism clvg. Dcmp-HCl (silica residue). Scndry in cavities, commonly with zeolites, etc.
X					$4\frac{1}{2}$–5	Pectolite $NaCa_2Si_3O_8(OH)$	Trcl	2.86–2.90	Radiating aggs. 2 perf clvgs. Most fluo-orange (lw). With zeolites in cavities.
X	X			X	$4\frac{1}{2}$–5	Wollastonite $CaSiO_3$	Trcl	2.87–3.09	Tblr, msv. Clvgs-84°+96°. Dcmp-HCl. Fluo-yel-orange. Cntct meta-ls.
X	X		X	X	$4\frac{1}{2}$–5	Scheelite $CaWO_4$	Tetr	6.08–6.12	Dipyramidal, msv, grnlr. Vitr. Clvgs. Unevn. Dcmp-HCl, HNO_3. Fluo-blue (X-ray, sw ultraviolet-, + cathode rays), thermolum. Hi-t vns, pegs, greissens, cntct meta rks.
X			X		5	Natrolite $Na_2Al_2Si_3O_{10}\cdot2H_2O$	Orth	2.20–2.26	Fibr grp. Sq needles. Prism clvg. Gelat. Pyroelectric. Zeol-assoc.
X			X		5	Scolecite $CaAl_2Si_3O_{10}\cdot3H_2O$	Mncl	2.25–2.29	Fibr grp. Thin striated prisms. Prism clvg. Gelat. Zeol-assoc, also hydrothermal and in metamorphosed calcareous rks.

X	X	X	X	X	5	Apatite ~$Ca_5(PO_4)_3(F,Cl,OH)$	Hexa	2.90–3.20	Group name. Most types indistinguishable megascopically. Hex prisms—commonly rounded, msv, grnlr. Vitr. Sol-acids. Some fluo-yel-orange. Ig rks, vns, pegs.
X	X			X	5	Hemimorphite $Zn_4Si_2O_7(OH)_2 \cdot H_2O$	Orth	3.45	Hemimorphic xls, sheaf-lk, mammillary, etc. 1 perf + 1 poor clvg. Gelat. Pyroelectric. Some fluo-pale orange (lw). Zn-deps.
X			X	X	5	Titanite (= sphene) $CaTiSiO_5$	Mncl	3.45–3.55	Double wedge-shaped xls, msv, grnlr. Admn. Alters to leucoxene. Dcmp-H_2SO_4. Common accessory constituent of ig and meta rks.
X		X	X		$5\text{–}5\frac{1}{2}$	Thompsonite $NaCa_2Al_5Si_5O_{20} \cdot 6H_2O$	Orth	2.10–2.39	Fibr grp. Radiated spherical masses. Gelat. Pyroelectric. Zeol-assoc.
X		X	X	X	$5\text{–}5\frac{1}{2}$	Datolite $CaBSiO_4(OH)$	Mncl	2.96–3.00	Tinted wht-colorless. Glassy/pearly. Gelat. Scndry with zeolites-cavities.
	X	X	X		$5\text{–}5\frac{1}{2}$	Goethite alpha-$FeO(OH)$	Orth	3.30–4.90	Submet. Yellowish strk. Msv, fibr, etc. Silky-admn. Sol-HCl. Wxing deps.
X		X	X	X	$5\text{–}5\frac{1}{2}$	Monazite $(Ce,La,Nd,Th)PO_4$	Mncl	4.60–5.40	Tblr-striated. Waxy-admn. Clvgs. Unevn. Dcmp-acids (slow). Access in granitic ig + gneissic rks, pegs, detrital sed rks.
X	X	X	X	X	5–6	Scapolite $(Na,Ca)_4[(Al,Si)_4O_8]_3\text{–}(Cl,CO_3)$	Tetr	2.50–2.78	Group name. Most types indistinguishable megascopically but S.G. may be indicative. Long, striated prisms + aggs of coarse xls. Subconch. Poor clvg giving irregular, striated-appearing surfaces. Marialite is insol; meionite is dcmp-HCl. Most fluo-yel-red (ultraviolet, lw). Cntct meta calc rks, pegs.

Table II–2 Minerals with Nonmetallic Lusters

Colorless/White	Gray/Black	Red/Orange	Yellow/Brown	Green/Blue/Purple	Hardness	Name, Formula	Crystal System	Specific Gravity	Remarks
X	X			X	5–6	Turquoise $CuAl_6(PO_4)_4(OH)_8 \cdot 4H_2O$	Trcl	2.60–2.84	Xls-rare, crusts, stalac. Waxy. 2 clvgs. Sol-HCl (difficultly). Wxed aluminous rks (especially in arid areas).
X	X			X	5–6	Hornblende (an amphibole) $Ca(Mg,Fe)_4Al(Si_7Al)O_{22}^{-}(OH,F)_2$	Mncl	3.00–3.40	Blk-greenish blk. Common amphibole in ig rks.
			X	X	5–6	Bronzite (an orthopyroxene) Mg-Fe silicate	Orth	3.30–3.45	Submet. (Bronzy). Fs12-30. In basic and ultrabasic ig rks, some metas.
X				X	5–6	Tremolite-actinolite (an amphibole) $Ca_2(Mg,Fe)_5Si_8O_{22}(OH)_2$	Mncl	2.98–3.46	Meta Si-bearing ls and dolo, lo-grade meta ultrabasic ig rks.
X	X		X	X	5–6	Amphibole Complex hydrous silicates of Mg + Fe ± Ca or Na (Al, etc.)	Mncls +Orth	2.85–3.57	Group name. Several end and other members with names unsettled for some. Many macroscopically indistinguishable but some colors and occurrences are indicative. Orth and mncl series. Long slender xls. Prism clvg at

	Hardness	Mineral	System	G	Remarks
X XX	5–6	Hypersthene (an orthopyroxene) Mg-Fe silicate	Orth	3.43–3.60	56°+124°. Sol-HF (slowly). Alters to biotite +/or chlorite. Ig + meta rks, vns. Fs30-50. Basic (e.g., norite) and ultrabasic ig rks, some meta rks.
X X X	5–6	Hematite alpha-Fe_2O_3	Hexa	5.26	Met-earthy. Strk-red. Plty, grnlr. Sol-HCl. Sed Fe-fms, rare in vns.
X X	5–6	Uraninite UO_2	Isom	10.63	Submet. Opq. Octah, etc., msv. Grsy-dull. Rdactv. Pegs, hi-mod-t vns.
X X	5–6½	Allanite $(Ce,Ca,Y)_2(Al,Fe)Si_3-O_{12}(OH)$	Mncl	2.80–4.20	Subtranslucent. Tblr, acicular, msv. Pitchy-resin. Subconch. Rdactv. Commonly metamict. Gelat. Access in ig rks, pegs.
X X X X X	5–7	Pyroxene Complex silicates	Mncl +Orth	2.96–3.96	Group name. Several end and other members with names unsettled for some. Many megascopically indistinguishable; occurrence may be indicative. Orth + mncl series. Short prisms. Prism clvg at 87° + 92°. Alters to amphibole. Ig + meta rks.
X X	5–7	Olivine $(Mg,Fe)_2SiO_4$	Orth	3.22–4.39	Group name for mins of fayalite-forsterite and fayalite-tephroite series, and monticellite and glaucochroite. Grnlr masses, dissem. Conch. Gelat. Commonly altered to serpentine. Basic and ultrabasic ig rks, cntct meta dolomitic ls, meteorites.
X X	5½	Analcime $NaAlSi_2O_6 \cdot H_2O$	Isom	2.24–2.29	Trapezohedra, msv, grnlr. Poor cubic clvg. Gelat. Amygdules-basalt.

Table II-2 Minerals with Nonmetallic Lusters

Colorless/White	Gray/Black	Red/Orange	Yellow/Brown	Green/Blue/Purple	Hardness	Name, Formula	Crystal System	Specific Gravity	Remarks
	X			X	5½	Augite (a clinopyroxene) $(Ca,Na)(Mg,Fe,Al,Ti)(Si,Al)_2O_6$	Mncl	3.74–3.85	The common pyroxene of subalkalic ig rks (e.g., gabbros).
X	X			X	5½–6	Sodalite $Na_8Al_6Si_6O_{24}Cl_2$	Isom	2.27–2.33	Dodec, msv. Poor dodec clvg. Gelat. Some fluo-yel-orange. In nepheline-bearing rks. Dissolve in HNO_3, evaporate solution slowly, halite remains.
X	X				5½–6	Nepheline $(Na,K)AlSiO_4$	Hexa	2.56–2.67	Tblr, prism. Vitr-grsy. Good prism, poor basal clvgs. Gelat. Fluo-orange. Distinguished from kalsilite and kaliophilite by X-ray. Na-rich alk ig and nearby meta rks.
X	X				5½–6	Anthophyllite (an amphibole) $(Mg,Fe)_7Si_8O_{22}(OH)_2$	Orth	2.85–3.57	With cordierite in gneisses, with talc in meta ultrabasic ig rks.
X	X		X	X	5½–6	Amblygonite $(Li,Na)Al(PO_4)(F,OH)$	Trcl	3.11	Equant, large masses. Vitr, pearly. Clvgs. Sol-acids (difficultly). In pegmatites.

H	Mineral (and notes) / Formula	System	G	Remarks
$5\frac{1}{2}$–$6\frac{1}{2}$	Opal $SiO_2 \cdot nH_2O$	Amorphous	1.80–2.25	Resin-pearly. Msv, mammillary crusts. Conch. Sol-HF and caustic alks. Some fluo-yel-green. Incrustations, in cavities (esp in volcanic rks).
$5\frac{1}{2}$–$6\frac{1}{2}$	Diopside (a clinopyroxene) $CaMgSi_2O_6$	Mncl	3.22–3.38	Some light ones fluo-blue. Wht-pale green. Meta calc-sil rks.
$5\frac{1}{2}$–$6\frac{1}{2}$	Rhodonite $(Mn,Fe,Mg,Ca)SiO_3$	Trcl	3.57–3.76	Tblr (commonly rough), msv. Alters readily to blk Mn-oxides.
$5\frac{1}{2}$–$6\frac{1}{2}$	Magnetite $FeFe_2O_3$	Isom	5.18	Submet. Magnetite series of spinel group. Blk strk. Octah, grnlr. Octah parting. Difficultly sol-HCl. Strongly magnetic. Wdsprd.
6	Sanidine (a feldspar) $KAlSi_3O_8$	Mncl	2.56–2.62	Tblr. Glassy. Optically distinct. Volcanic rks.
6	Lawsonite $CaAl_2Si_2O_7(OH)_2 \cdot H_2O$	Orth	3.05–3.10	Prism, tblr. Clvg-2 perf, 1 poor. Lo-grade (e.g., glaucophane) schists.
6	Glaucophane (an amphibole) $Na_2(Mg,Fe)_3Al_2Si_8O_{22}(OH)_2$	Mncl	3.08–3.30	With lawsonite, etc. In schists + gneisses (commonly Na-rich ones).
6	Humite $(Mg,Fe)_7(SiO_4)_3(F,OH)_2$	Orth	3.20–3.32	Group name, also applied to one species of group. Irregularly shaped grns. Vitr-resin. Gelat. Similar to olivine. Cntct metamorphosed dolomitic ls.
6	Zoisite $Ca_2Al_3(SiO_4)_3(OH)$	Orth	3.15–3.37	Distinguished from clinozoisite by optical or X-ray anal. Thulite is pink variety.

Table II-2 Minerals with Nonmetallic Lusters

Colorless/White	Gray/Black	Red/Orange	Yellow/Brown	Green/Blue/Purple	Hardness	Name, Formula	Crystal System	Specific Gravity	Remarks
X				X	6	Jadeite (a clinopyroxene) Na(Al,Fe)Si$_2$O$_6$	Mncl	3.24–3.43	Relatively uncommon, typically with albite, hi-t and also lo-grade meta rks.
X			X	X	6	Epidote Ca$_2$(Al,Fe)$_3$ (SiO$_4$)$_3$(OH)	Mncl	3.38–3.49	Group name. No generally accepted nomenclature for group members. Closely related to zoisite and allanite. Pistachio-blackish green-brownish. Long, thin, grooved xls, grnlr masses. Vns, other cavities, regional meta + ig rks.
	X		X	X	6	Aegirine/acmite (a clinopyroxene) NaFeSi$_2$O$_6$	Mncl	3.40–3.55	Typical pyroxene of alkalic rks- especially syenitic ones.
	X		X		6	Pigeonite (a clinopyroxene) (Mg,Fe,Ca)(Mg,Fe) Si$_2$O$_6$	Mncl	3.30–3.46	Common pyroxene of many fine-grained gabbros and basalts.
X	X	X			6	Columbite FeNb$_2$O$_6$	Orth	5.15–5.25	Forms series with tantalite. Submet. Red-blk strk. Tarn-irid. Tblr. prism. 2 clvgs. Brittle. Granitic pegs.

			H	Mineral	System	G	Remarks
X	X		$6-6\frac{1}{2}$	Albite (a plagioclase) $NaAlSi_3O_8$	Trcl	2.57–2.63	An0–10. See feldspar. Ig, meta rks, pegs. Moonstone and cleavelandite are varieties.
X	X	X X	$6-6\frac{1}{2}$	Orthoclase (a feldspar) $KAlSi_3O_8$	Trcl	2.55–2.63	Distinguished from microcline optically. Plutonic ig rks, pegs.
X	X	X X	$6-6\frac{1}{2}$	Microcline (a feldspar) $KAlSi_3O_8$	Trcl	2.56–2.63	Distinguished from orthoclase optically. Plutonic ig rks, pegs.
X	X	X X	$6-6\frac{1}{2}$	Perthite (feldspar)	Trcl	2.56–2.65	Megascopically, microscopically, or submicroscopically interdigitated microcline or orthoclase and plagioclase (typically albite).
X	X	X	$6-6\frac{1}{2}$	Oligoclase (a plagioclase)	Trcl	2.62–2.67	An10–30. See feldspar. Plutonic ig rks, few pegs. Sunstone is var.
X	X		$6-6\frac{1}{2}$	Andesine (a plagioclase)	Trcl	2.64–2.69	An30–50. See feldspar. Intermediate Si-content rks.
X	X		$6-6\frac{1}{2}$	Labradorite (a plagioclase)	Trcl	2.68–2.72	An50–70. Bluish reflections. See feldspar. Basic ig rks.
X	X	X X	$6-6\frac{1}{2}$	Feldspar (plagioclase series: $NaAlSi_3O_8$–$CaAlSi_2O_8$; potassium feldspar: $KAlSi_3O_8$)	Trcl	2.55–3.39	Includes plagioclase (Na-Ca) series, alkali (K-Na) and Ba feldspars. Most easily distinguished by nonmegascopic means. Monoclinic ones have 2 clvgs at 90° and simple (if any) twinning; triclinic ones have 2 clvgs up to 4° off 90° and typically have polysynthetic twinning (giving striated appearance on some clvg surfaces of plagioclase). Pegmatites; ig, meta, and some sed rks.

Table II–2 Minerals with Nonmetallic Lusters

Hardness	Name, Formula	Crystal System	Specific Gravity	Remarks	Colorless/White	Gray/Black	Red/Orange	Yellow/Brown	Green/Blue/Purple
6–6½	Prehnite $Ca_2Al_2Si_3O_{10}(OH)_2$	Orth	2.90–2.95	Rosettes-tblr xls. 1 clvg. Sol-HCl (slow). Cavities (with zeolite).	X				X
6–6½	Rutile TiO_2	Tetr	4.21–4.25	Submet. Prism-striated. Admn. Poor clvgs. Vns, access-meta + ig rks.		X	X	X	
6–6½	Tantalite $(Fe,Mn)Ta_2O_6$	Orth	7.90–8.00	Forms series with columbite. Submet. Pegs.		X	X	X	
6–7	Vesuvianite $Ca_{10}Mg_2Al_4(SiO_4)_5(Si_2O_7)_2(OH)_4$	Tetr	3.33–3.43	Prism, msv. 1 poor clvg. Subconch. Attkd-HCl. Meta-esp cntct ls.		X		X	X
6–7	Cassiterite SnO_2	Tetr	6.99	Submet. Radial concretionary masses. Admn-dull. Hi-t vns. greisens.		X	X	X	
6–7½	Pyrope (a garnet) $Mg_3Al_2(SiO_4)_3$	Isom	3.58	Mg-Al garnet. Xls are rare. Alters to kelyphite. Ultrabasic ig rk.		X	X	X	
6–7½	Grossular (a garnet) $Ca_3Al_2(SiO_4)_3$	Isom	3.59	Ca-Al garnet. Equant xls. Contact and regional meta calc rks.	X				X

	H	Mineral	Crystal system	G	Remarks
X X X	$6-7\frac{1}{2}$	Andradite (a garnet) $Ca_3Fe_2(SiO_4)_3$	Isom	3.86	Ca-Fe, Ti garnet. F-mag glbl. Cntct meta calc rks, alk igs.
X	$6-7\frac{1}{2}$	Uvarovite (a garnet) $Ca_3Cr_2(SiO_4)_3$	Isom	3.90	Ca-Cr garnet. Dodec xls. With chromite in serpentinites, skarns.
X X X	$6-7\frac{1}{2}$	Spessartine (a garnet) $Mn_3Al_2(SiO_4)_3$	Isom	4.19	Mn-Al garnet. Equant xls. Mn-rich meta rks + vns, pegs.
X X X	$6-7\frac{1}{2}$	Almandine (a garnet) $Fe_3Al_2(SiO_4)_3$	Isom	4.32	Fe-Al garnet. F-mag glbl. Meta-argillaceous rks, rare-ig rks, pegs.
X X X X X	$6-7\frac{1}{2}$	Garnet Silicates of Fe,Mg,Ca, etc.	Isom	3.58–4.32	Group name. Composition determines name. Distinguished by S.G. + nonmegascopic means. Equant xls, msv. Vitr-dull. Dodec parting. Slightly- to in-sol in HF.
X X X X X	$6\frac{1}{2}$	Agate SiO_2	Hexa	2.57–2.64	Banded or variegated chalcedony (microxline qtz). Nodules in basalt.
X X X X	$6\frac{1}{2}$	Chalcedony SiO_2	Hexa	2.57–2.64	Microxline and microfibr qtz. Mammillary. Lo-t vns and other cavities.
X X	$6\frac{1}{2}$	Carnelian SiO_2	Hexa	2.57–2.64	Red to brwnish red chalcedony (microxline qtz).
X X X X X	$6\frac{1}{2}$	Clinozoisite $Ca_2Al_3(SiO_4)_3(OH)$	Mncl	3.21–3.38	Ca-Al end member of epidote group.
X X X	$6\frac{1}{2}-7$	Jasper SiO_2	Hexa	2.57–2.65	Subtranslucent. Some is variegated or banded (chalcedony). Microgrmlr. Cavities.

Table II–2 Minerals with Nonmetallic Lusters

Colorless/White	Gray/Black	Red/Orange	Yellow/Brown	Green/Blue/Purple	Hardness	Name, Formula	Crystal System	Specific Gravity	Remarks
X	X		X	X	6½–7	Spodumene (a clinopyroxene) LiAlSi$_2$O$_6$	Mncl	3.03–3.22	Commonly fluo (and phosphoresces)-orange. Thermolum. Li-bearing pegs.
X	X		X		6½–7½	Andalusite Al$_2$SiO$_5$	Orth	3.13–3.16	Blunt, nearly sq. prisms. Commonly altered. Cntct and regional meta.
X			X		6½–7½	Sillimanite Al$_2$SiO$_5$	Orth	3.23–3.27	Prismatic xls, fibr masses. Vitr. 1 clvg. Hi-grade meta rks.
X					7	Tridymite SiO$_2$	Orth	2.25–2.27	Plty microxls. Vitr-pearly. Conch. Sol-boiling NaCO$_3$. Silicic vols.
X	X	X	X	X	7	Quartz SiO$_2$	Hexa	2.65	Singly and doubly terminated hex prisms striated perpendicular to length (lo-t), bipyramids (hi-t), msv. Vitr. Conch. Sol-HF. Wdsprd in most rk types + vns. (Pink = rose quartz; yellow = citrine; purple = amethyst; brownish-gray = cairngrom = smoky quartz.)
X	X			X	7	Cordierite Mg$_2$Al$_4$Si$_5$O$_{18}$	Orth	2.53–2.78	Color differs with direction. Dissem grns. 1 good clvg. Meta rks.
X	X	X	X	X	7	Tourmaline Complex borosilicates	Hexa	3.03–3.25	Group name. Varieties commonly distinguishable megascopically on basis of color(s) (e.g., black = schorl; brown = dravite; pink, green,

colorless=elbaite). Lengthwise striated prismatic xls with cross sections that resemble spherical triangles. Some xls exhibit both lengthwise + concentric color zoning. Vitr-resin. Brtl. Electrically charged on heating and cooling. Mg-rich varieties fluo-yel (ultraviolet-sw). Pegs, hi-t vns, ig and meta rks.

Props	H	Mineral / Formula	Syst	G	Remarks
X	7	Schorl $NaFe_3Al_6(BO_3)_3Si_6O_{18}(OH)_4$	Hexa	3.10–3.25	Na-Fe tourmaline. Typically blk. Ig rks, schists and gneisses.
X X	$7\frac{1}{2}$	Staurolite $(Fe,Mg,Zn)_2Al_9(Si,Al)_4O_{22}(OH)_2$	Orth	3.74–3.83	Prism-cruciform twins common. Fires to mag powder. Schist and gneiss.
X X X	$7\frac{1}{2}$	Zircon $ZrSiO_4$	Tetr	4.60–4.70	Terminated prisms. Admn. Some fluo-yel-orange. Access-ig, sands.
X X X	$7\frac{1}{2}$–8	Beryl $Be_3Al_2Si_6O_{18}$	Hexa	2.66–2.83	Grooved prism. Poor basal clvg. Some fluo weakly yel. Pegs.
X X X	$7\frac{1}{2}$–8	Spinel $MgAl_2O_4$	Isom	3.55	Group name. Some have distinctive colors or other hand-specimen props. Octah. Grnlr. Glassy. Conch. Some fluo-red to yel-green. In meta (esp calc rks), pegs, and placers.
X X X	8	Topaz $Al_2SiO_4(F,OH)_2$	Orth	3.49–3.57	Striated prisms. Perf basal clvg. Cntct zones, greisens, pegs.
X X X	9	Corundum Al_2O_3	Hexa	4.00–4.10	Steep pyramidal, prism. Rhomb and basal parting. Wdsprd in Si-deficient rks and placers.
X X X X	10	Diamond C	Isom	3.50–3.53	Octah. Admn-grsy. Brtl. Octah clvg. Ultrabasic igs and placers.

APPENDIX III

Rock Identification Tables

The rocks included in Tables III-1–III-3 are relatively common and/or widespread. The tables are slightly modified versions of those given to aid macroscopic identification of rocks in R. V. Dietrich and B. J. Skinner, *Rocks and Rock Minerals* (New York: Wiley, 1979). Several additional rocks are also described in that book.

Because only the common types of rocks are included on the tables, you will not be able to identify every rock by using them. Remember, too, that several rocks may fit between rocks given in the table—for example, a sandy conglomerate (see Figure 3–14).

As is noted in *Rocks and Rock Minerals*:

> There have been many attempts to outline the mental processes used in identifying and naming a rock. The attempts have rarely been useful. In part this is so because the procedures outlined are seldom the ones actually used by professional geologists. Most professionals, in fact, do not consciously go through any set procedure—rather, they identify rocks much as they recognize their friends. Nonetheless, our experience indicates that a good first step is to use charts like [Tables III-1, III-2, III-3] and a procedure as outlined . . . below. By so doing, most persons will fairly quickly build up the necessary background so that they, too, can then identify and name most rocks in a more or less routine manner.

Tables III-1 and III-2 show rocks whose main constituents are present as macroscopically discernible grains or which may be readily identified by a few simple tests. Table III-3 is for rocks whose main constituents are too small for macroscopic identification.

All the tables have two main subdivisions based on hardness, and additional subdivisions based on rock genesis. The two hardness categories separate Table III-1 from III-2, and are indicated by a heavy horizontal line in Table III-3. Genetic categories are separated by single vertical lines in all tables. In Tables III-1 and III-2 there are three additional subdivisions, separated by double vertical lines, that are based on some overall textural features. Finally there are remarks about some of the individual rocks.

The hardness categories are (1) rocks that will scratch a hammer and (2) rocks that will not. The hammer is used as the standard because most collectors carry one while they are in the field. In general, it is best to try to scratch the hammer with the rock, not the reverse. Also, it is wise to use several corners of a specimen so that some rare constituent present on, for example, the first corner checked doesn't mislead you. In addition, with rocks that tend to break apart between grains, it may be necessary to rub some of the small pieces on the hammer and, in some instances, to examine the area rubbed with a hand lens to see whether it has or has not been scratched.

The genetic categories are described in Chapter 3. In many cases, relationships you will see in the field will permit you to use these divisions to advantage.

Here are two examples of how you may use the tables.

First Specimen:

1. Grains: macroscopic—use Tables III-1 and III-2.
2. Hardness: less than hammer—use Table III-2.
3. Texture: interlocking—use left third of the tables.
4. Rock is:
    a. Rock salt, if it has a salty taste.
    b. Rock gypsum, if it can be scratched with a fingernail.
    c. Limestone or calcite marble, if it effervesces briskly with dilute HCl.
    d. Dolostone or dolomite marble, if it effervesces slowly with dilute HCl.
    e. Rock anhydrite, if it does none of these.

Second Specimen:

1. Grains: submacroscopic—use Table III-3.
2. Hardness: greater than hammer—use top section of table.
3. Rock is one of the following: felsite, basalt, obsidian, pumice, ash tuff, chert, shale, diatomite, slate, phyllite, greenstone, or mylonite, each of which may be distinguished on the basis of the remarks—for example, note the color, luster, and translucent quality of obsidian.

Table III–1 Identification of Rocks with Macroscopically Discernible Grains; Hardness > Hammer.*

Interlocking Grains (Figure 3–2)			Foliated (Figure 3–6)	Fragmental and/or Layered, (Figures 3–3 and 3–4)		
Igneous	Sedimentary/ Diagenetic	Metamorphic	Metamorphic	Igneous/ Pyroclastic	Sedimentary/ Diagenetic	Metamorphic
After constituent minerals are identified, see Fig. 3.9 and Tbl. 3-1. Compositions below are typical, *not* inclusive. Q = quartz, Alk = alkali feldspar, Plag = plagioclase, M = mafics. *Granitoid,* Q, 20–60%; Alk > = < Plag; M,		All rocks, except phyllite, in the next ("Foliated") column also may consist of interlocking grains. *Quartzite* H-7; vitreous, conchoidal fracture.	Quartz is common in all of these except amphibolite. *Gneiss* Streaked or banded; chiefly granular grains. *Amphibolite* Dark gray to greenish black; medium grained; lacks quartz *Schist* Enough platy	The first four rocks are named on the basis of size of the clasts that make up 50 or more percent of the volume of the rock. *Agglomerate* (bomb tuff) Clasts are bombs >64 mm. *Pyroclastic breccia* (Block tuff) Clasts are blocks >64	Rocks consist chiefly of clasts with mean diameters of noted lengths. *Conglomerate* Rounded clasts > 2 mm. *Sedimentary breccia* Angular clasts > 2 mm. *Sandstone* 1/16–2 mm; quartz > 75% *Arkose* 1/16–2 mm;	*Quartzite* Chiefly quartz; conchoidal fracture; may be sedimentary—see p. 75. *Metaconglomerate* Pebbles, cobbles, and boulders in a metaquartzite matrix.

feldspar >
25%;
resembles
granite
Graywacke
1/16–2mm;
mafics ±
rock clasts >
25%; clay
odor when
damp.
Siltstone
1/256–1/16
mm; gritty
between
teeth.

mm.
Lapilli tuff
Clasts 2–64
mm.
Ash tuff
Clasts <2 mm;
rough to
touch.

constituents
to give good
foliation.
Phyllite
Very fine
grained;
foliation
surfaces have
glossy sheen;
commonly
corrugated.

10–40%.
Syenitoid
Q < 20%; Alk,
up to 100%;
M < 45%.
Dioritoid
Q < 5%; light
Plag > 50%;
M < 50%
(hornblende).
Gabbroid
Q < 5%; dark
Plag < 50%;
M > 50%
(pyroxene).
Anorthosite
Bluish-gray
Plag,
90–100%.
Ultramafites
One or more
mafic
minerals,
90–100%.

*Revised after Dietrich, R. V., and Skinner, B. J., *Rocks and Rock Minerals*. New York: Wiley, 1979, pp. 302–303.

Table III–2 Identification of Rocks with Macroscopically Discernible Grains; Hardness < Hammer.*

Interlocking Grains, Figure 3–2		Foliated, Figure 3–6	Fragmental and/or Layered, Figures 3–3 and 3–4
Sedimentary/Diagenetic	Metamorphic	Metamorphic	Sedimentary/Diagenetic
Rock salt Has salty taste. *Rock gypsum* Can be scratched with fingernail. *Rock anhydrite* H—3½ *Limestone* Brisk effervescence with dilute HCl. *Dolostone* Slow smoldering effervescence with dilute HCl.	*Marble* May be calcitic to dolomitic; commonly contains disseminated graphite and/or silicate minerals; with HCl effervesces as noted in column at left.	*Talc Schist* Can be scratched with fingernail; soapy feel; is foliated. *Soapstone* Like talc schist but nonfoliated. *Serpentinite* Typically green; H—2½–3½; waxy to greasy appearance.	*Clastic limestone* Contains fragments including intraclasts, fossils, oolites, and/or pellets; effervesces briskly with dilute HCl. *Dolostone* As above but effervesces slowly with dilute HCl.

*Revised after Dietrich, R. V., and Skinner, B. J. *Rocks and Rock Minerals*. New York: Wiley, p. 304.

Table III–3 Identification of Rocks That Are Microcrystalline or Glassy.

Igneous	Pyroclastic	Sedimentary/ Diagenetic	Metamorphic
Hardness > Hammer:	Ash tuff, p. 163 Consolidated ash; rough feel.	Chert, p. 217 Porcelaneous luster; conchoidal fracture.	Slate, p. 265 Parallel rock cleavage that may be at an angle to bedding.
Felsite, p. 147 Light-colored (p. 147); stony appearance.			
Basalt, p. 150 Dark gray to greenish black; commonly vesicular or amygdaloidal.		Shale, p. 200 Disaggregates easily; is fissile; clay odor when damp.	Phyllite, p. 265 Glossy sheen on foliation surfaces; commonly corrugated.
Obsidian, p. 151 Glassy; dark gray, brown or streaked; translucent in thin pieces; vitreous luster.			Greenstone, p. 262 Olive green color; dull luster; subconchoidal fracture.
Pumice, p. 157 Glassy; frothlike.			Mylonite, p. 247 See below.

Igneous	Pyroclastic	Sedimentary/ Diagenetic	Metamorphic
Hardness < Hammer		*Coal* Black color and streak; dull and/or bright luster; brittle. *Rock gypsum* Can be scratched with fingernail. *Rock anhydrite* H—3½. *Limestone* Brisk effervescence with dilute HCl. *Dolostone* Slow smoldering effervescence with dilute HCl. *Claystone* Clay odor when damp; smooth feel; sticks to tongue; no effervescence with acid.	*Mylonite* Smeared-out appearance; commonly includes sheared fragments.

APPENDIX IV

Geologic Time

An appreciation of the immensity of geologic time is fundamental to an understanding of the history of life and of the role fossils play in evolutionary theory.

What, then, is geologic time, and how is it measured? *Relative geologic time* is based on the sequential order of events, from earliest to latest, without knowing when, in years, the events took place. Relative time is determined mainly through the use of fossils. *Absolute geologic time,* on the other hand, is based on the number of years before the present that an event took place. Absolute time is determined by radiometric dating of rocks.

The study of fossils and the principles of stratigraphy, given in Chapter 4, led to the development of the currently used Geologic Time Scale (Table IV-1). This time scale was originally a relative-time "calendar" based on events preserved in the fossil record. The discovery of radioactive "clocks" in rocks enabled geologists to determine the number of years before the present that events took place in the geologic past and thus provided the absolute-time numbers given on the chart.

The early development of the scale was closely linked to the study of the history of life, because the appearance and extinction of animals and plants were used to mark the limits of many of the subdivisions on the scale. It was recognized during the 1800s that the earth's sedimentary rocks could be subdivided into relative time units on the basis of the fossils they contained. It was also observed that there was an order to the appearance and disappearance of fossils in the rock column, and that this order was the same for different rocks representing different environments, even in widely separated localities. These observations formed the basis for the principles of superposition and of faunal and floral succession, both of which are noted in Chapter 4.

Table IV–1 Geologic Time Scale (Number Gives Dates in Millions of Years Before the Present).

Eras	Periods	Epochs	
Cenozoic	Quaternary	Recent	0.01
		Pleistocene	1.5
	Tertiary	Pliocene	7
		Miocene	26
		Oligocene	38
		Eocene	54
		Paleocene	
		65	
Mesozoic	Cretaceous	136	
	Jurassic	190	
Paleozoic	Triassic	225	
	Permian	280	
	Pennsylvanian	320	
	Mississippian	345	
	Devonian	395	
	Silurian	430	
	Ordovician	500	
	Cambrian	570	
Precambrian time			

These principles are especially important because nowhere in the world is there exposed a complete sequence of sedimentary rocks representing the time from the earth's formation to the present day. Therefore, we must—on the basis of the two principles just mentioned—correlate widely separated sedimentary rock exposures in order to reconstruct the geologic history of the earth (see Figure 4–12).

By placing geological events in a sequential order, and by correlating sedimentary rock exposures on the basis of their contained fossils, the geologic time scale gradually evolved. Thus, it is based on major geologic and biologic events that have been preserved in the rock record. That is to say, the geologic time scale was originally a relative time scale, with events placed in their sequential order, but with no indication of how long before the present the events took place.

How, then, were absolute time values, representing the actual number of years before the present, determined? The answer is by *radiometric dating of rocks*.

Elements such as uranium, thorium, rubidium, and potassium have radioactive isotopes that are unstable and spontaneously decay to the stable form of another element. For example, radioactive uranium 238 decays, through several steps, to stable lead 206. This decay and transformation takes place at a constant rate, regardless of the external or internal forces to which the element is subjected. And the decay rate can be measured with great precision. Consequently, by knowing how long it takes *half* of the original radioactive material to decay to its stable daughter (a period of time known as the isotope's *half-life*), and by comparing the amount of the parent radioactive element actually present with the amount of the stable daughter element, one can calculate how old the containing rock is.

Radiometric dating, then, allows us to determine how long ago an event recorded in the geological record took place and, most important, it allows us to assign dates to the geologic time scale.

The geologic time scale is divided into three progressively smaller divisions of time: *eras, periods, and epochs*. It should be noted that the divisions of the geologic time scale are not equal. This is because the boundaries were originally based on geologic and biologic events.

To gain an appreciation of geologic time and when major evolutionary events took place, compare the geologic history of the earth with one of our present-day calendar years (Table IV-2).

Table IV–2 Geological History of the Earth, Compressed Into One Calendar Year. The Events Are Placed in Stratigraphic Order, with the First Event on the Bottom and the Last Event at the Top.

Years Before Present	Event	Days Since January 1	Date and Time
35	World War II ends	364.9	December 31, 11:59:57.6 P.M.
10,000	Ice age ends	364.9	December 31, 11:58:49.9 P.M.
1.75 million	First human appears	364.9	December 31, 7:35:36.0 P.M.
3 million	Ice age begins	364.7	December 31, 6:09:36.0 P.M.
55 million	Horses first appear	360.5	December 27, 12:56:00.0 P.M.
65 million	Dinosaurs become extinct	359.7	December 26, 5:28:00.0 P.M.
130 million	First flowering plants	354.5	December 21, 10:56:00.0 P.M.
160 million	First birds	352.0	December 19, 5:19:59.9 P.M.
215 million	First mammals	347.6	December 14, 1:27:59.9 P.M.
250 million	Dinosaurs first appear	344.7	December 11, 4:19:59.9 P.M.
360 million	First amphibians	335.8	December 2, 7:12:00 P.M.
420 million	First land plants	330.9	November 27, 10:23:59.9 P.M.
470 million	First fish	326.9	November 23, 9:40:00 P.M.
700 million	First metazoans	308.2	November 5, 5:19:59.9 P.M.
3.5 billion	Oldest fossils	105.4	April 16, 10:39:59.9 A.M.
4.5 billion	Earth formed	–0–	January 1, 12:00 midnight

APPENDIX V

Divisions of the Animal and Plant World

Most fossil collectors are amazed when they first realize the tremendous variety of fossils that are known, and consequently they may feel overwhelmed when they start to try to identify what they have. Nonetheless, we can assure you that it usually is relatively easy to identify a fossil down to its class, and, by using a book or two about that particular group, to identify the fossil down to its species.

We suggest that you briefly review the section on how plants and animals are classified (pages 82–84) before proceeding through this appendix. Remember, the classification scheme is a hierarchical system of grouping similar objects together, going from the largest grouping with very general characteristics, down to the smallest, with many specific characteristics. When you remember that relationship, the task of identifying your fossils will not seem so formidable.

This appendix is intended as a general guide to the major animal and plant phyla and classes. After locating a fossil here, you can turn to a specific section of a more specialized text for specific identification. This appendix is not meant to serve as a reference for identifying fossils down to their species.

First we give a brief description of where marine animals live and how they live. Then, we provide a brief summary of each major group of invertebrate and vertebrate animals, and the major plant groups.

The marine environment can be divided into (1) the sea floor and (2) the water above the floor. The sea floor environment is called the *benthonic* zone; the ocean itself is called the *pelagic* zone. The portion of the water zone that sunlight can penetrate is called the *photic* zone; it ranges in depth from nearly zero to 200 meters.

Organisms that live in the pelagic zone can be divided into two groups. The first is *plankton,* which are floaters. Most of them are passive and go where the currents take them. Plankton life can be divided into the *phytoplankton,* the plants, and the *zooplankton*, the animals.

Most phytoplankton are microscopic plants, such as diatoms, dinoflagellates, and algae, that live in the photic zone where they can use sunlight to produce their food. Seaweeds, which are large marine plants, can be carried around by currents and waves and are also considered plankton.

Most zooplankton, which also are generally small (typically microscopic), feed on phytoplankton. Because they feed on phytoplankton, most zooplankton also live in the photic zone. The zooplankton consist of microscopic forms, such as foraminifera and radiolarians, as well as larger animals, such as jellyfish.

The *nekton* comprise a second group of animals that live in ocean water; they are swimmers. Nearly all nektonic animals are vertebrates such as fish; The *cephalopods*—squids and octopi—are the only invertebrate nektonic animals. The vast majority of nektonic animals are active predators.

Benthonic plants and animals that live on the sea floor comprise the *epiflora* and *epifauna,* whereas those that live below the surface of the sea floor are called the *infauna*. The epifaunal and infaunal animals can be divided into those that move around, the *mobile* or *vagrant* forms, and those that are fixed in one place, the *sessile* forms. Examples of mobile animals are starfish, arthropods, and snails; examples of sessile animals are corals, brachiopods, and bryozoans.

The way they eat is also important, and provides us with yet another way of grouping animals.

Filter feeders or *suspension feeders* strain or filter microscopic plants and animals, as well as dissolved nutrients, from the water. The sessile benthonic organisms are all filter feeders and thus rely on the constant stream of organic material in the water for their food. The filter feeders employ a variety of methods for extracting their food from the water. Filter feeders include the sponges, brachiopods, corals, and pelecypods.

Herbivores eat only plants. Some gastropods and chitons are invertebrate herbivores; horses and cows are examples of vertebrate herbivores.

Carnivores are meat eaters. Starfish, cephalopods, and some gastropods are good examples of carnivorous invertebrates; lions, alligators, and sharks are vertebrate carnivores.

Omnivores eat both meat and plants. Humans are good examples of omnivores.

Detritus-deposit feeders eat the material on the sea floor and/or ingest and extract nutrients from sediment. The detritus-deposit feeders include some worms, some arthropods, and some echinoderms.

We can now define an organism by where it lives and how it eats. For example, a pelecypod is a benthonic, epifaunal, filter feeder; a cephalopod is a nektonic carnivore; and a worm is a benthonic, infaunal, detritus-deposit feeder.

Animals and plants interact with each other and are distributed in a community according to both biologic and physical factors. The way organisms interact defines a *food chain* that indicates the flow of energy from each level of the community to a higher level.

The simplest type of food chain consists of *producers* and *decomposers*. Simple plants, such as algae, produce their own food, whereas bacteria convert once-living things into compounds that can be used directly by other plants and animals.

Most food chains consist of *producers, consumers,* and *decomposers*. Examples of these range from a simple one consisting of algae (producers), filter feeders (consumers), and bacteria (decomposers), to more complex chains consisting of algae (producers), different types of consumers such as filter feeders being eaten by carnivores, and all being decomposed by bacteria.

Remember that in the living world there are complex interrelationships among virtually all organisms, so that every fossil collection represents, to some degree, a similar set of relationships that existed in the past.

Before proceeding to the discussion of the animal phyla, it would be helpful to review Table V-1 to see the diversity of the animal world.

Table V–1   Classification of Animals

Phylum, Subphylum	Superclass, Class	Common Names
Protozoa	Sarcodina	Foraminifera, radiolaria
Porifera		Sponges
Coelenterata	Scyphozoans	Jellyfish
	Hydrozoans	Stromatoporoids
	Anthozoans	Tabulate, rugose, and scleractinia corals
Bryozoa		Moss animals
Brachiopoda		Brachiopods
Mollusca	Amphinera	
	Scaphopoda	Tusk shells
	Gastropoda	Snails
	Pelecypoda	Clams, oysters, scallops
	Cephalopoda	Nautiloids, squid, octopus
Annelida		Worms
Arthropoda	Trilobita	Trilobites
	Meristomata	Horseshoe crabs, scorpions, spiders, eurypterids
	Crustacea	Shrimps, crabs, lobsters, copepods
	Ostracoda	Ostracodes
	Cirripedia	Barnacles
	Insecta	Insects
Echinodermata		
Pelmatozoa	Cystoidea	Cystoids
	Blastoidea	Blastoids
	Crinoidea	Crinoids or sea lillies
Eleutherozoa	Holothuroidea	Sea cucumbers
	Stelleroidea	Starfish, brittle stars
	Echinoidea	Sea urchins, heart urchins, sand dollars

(*continued on next page*)

Phylum, Subphylum	Superclass, Class	Common Names
Protochordata	Graptolithina	Graptolites
Chordata	Pisces	Fishes
	Tetrapoda	Amphibians, reptiles
	Aves	Birds
	Mammals	Horses, elephants, humans, among others

ANIMAL PHYLA

Phylum Protozoa: The protozoa are one-celled animals, some of which are soft-bodied (for example, the amoeba) and others of which produce a shell or a skeleton (for example, the foraminifera and radiolaria). Some protozoans are benthonic; others are planktonic. They have a geologic time range from Precambrian time through the Present. They are divided into four classes on the basis of their locomotion. Most paleontologists are concerned with only one class, the sarcodina.

The *sarcodina* are one-celled animals that possess pseudopodia (mobile extensions of the body) used in both feeding and locomotion. The two important orders of the sarcodina are the *foraminifera* and *radiolaria*.

The *foraminifera* are predominantly marine organisms (Figure V-1). They are typically microscopic (less than 1 millimeter) in size, although some forms can attain diameters of up to 100 millimeters (~ 4 inches). They either secrete a calcareous shell around their cytoplasm, or construct one of loose particles of material available on the sea bottom. The latter type of shell is referred to as an agglutinated shell or test. The majority of foraminifera are benthonic; these are

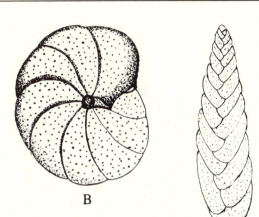

Figure V–1. Foraminifera. (A) *Globorotalia,* a planktonic species. (B) *Pullenia,* a benthonic species. (C) *Bolivina,* a benthonic species.

thought to be the first type to have evolved. The others are planktonic. Both groups are found at essentially all depths and in all the oceans throughout the world. The foraminifera, because of their small size, abundance, and short geologic ranges are one of the most useful fossil groups for age-dating. They are also important as rock-building agents—for example, the Great Pyramids of Egypt are constructed from foraminiferal limestone.

The *radiolarians* are exclusively planktonic marine organisms (Figure V–2). They construct an internal skeleton of silica and float at or near the surface of the ocean. They occur all over, but are especially abundant in warmer waters. When radiolarians die, their skeletons fall to the bottom and produce distinctive siliceous deposits known as radiolarian ooze. Former oozes are abundant in some rock formations, and are the parent material for certain chert deposits, such as the Franciscan Chert of California. The geologic time range of radiolarians is from Precambrian time through the Present. Although radiolarians are not as generally useful as foraminifers in stratigraphic studies, they have proved very useful for age-dating and correlation of several deep sea deposits, particularly in the Deep Sea Drilling Project sponsored by the National Science Foundation.

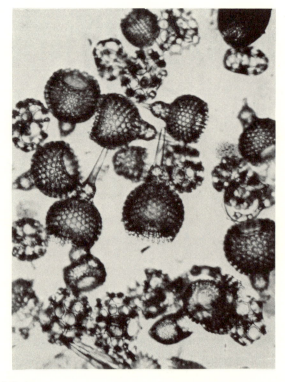

Figure V–2.   Radiolarians.

Phylum Porifera: The *porifera* or *sponges* (Figure V-3) constitute a step up in complexity from the protozoans. Simplest of the multicelled animals, they consist of three pore-bearing layers of cells, each of which has specialized functions. The cells are not, however, organized into definite tissue. The sponges have no internal organs, and no nervous, digestive, or circulatory system.

The sponges are benthonic filter feeders that occur in both fresh and marine waters. They range in size from less than 1 millimeter to two meters in greatest dimension. The sponges occur both as solitary individuals and as colonies, with colonial marine forms the most common.

The skeletons of sponges consist of internally secreted siliceous, calcareous, or organic *spicules* that are disseminated throughout the sponges' bodies, either as individual elements or fused together in lattice networks. The spicules (Figure V-3) are of diverse shape and composition; these differences are the basis for sponge classification.

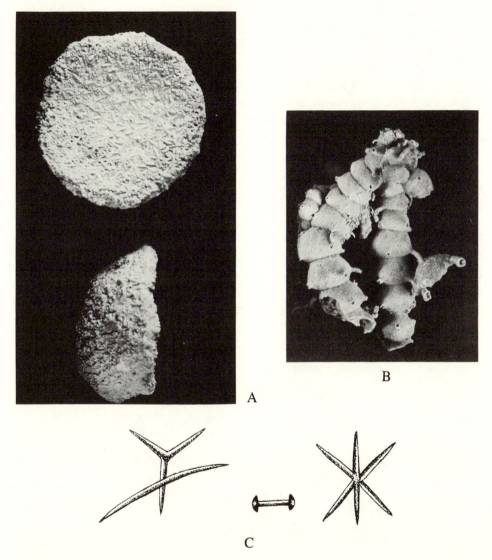

Figure V–3. Porifera. (A) *Astraeospongia meniscus*. (B) *Girtycoelia*. (C) Sponge spicules. (Photographs courtesy of Smithsonian Institution.)

Generally only the spicules are preserved, although the overall shapes and structures of some sponges have also been preserved in some instances.

The geologic time range of the sponges is from Precambrian time through the Present.

Phylum Coelenterata: The *coelenterates* are aquatic animals of diverse forms that have achieved the tissue level of organization. Except for a few freshwater forms, the coelenterates are all inhabitants of the sea. They are characterized by radial symmetry and a two-layered wall surrounding a body cavity in which digestion takes place. They lack respiratory, circulatory, excretory, and central nervous systems. Many of them also lack hard parts, but others secrete calcium carbonate to form skeletal structures that are readily fossilized.

Most paleontologists and zoologists recognize three classes of coelenterates: the *scyphozoans,* the *hydrozoans,* and the *anthozoans.*

The *scyphozoans,* or *jellyfish,* lack hard parts and consequently are fossilized only under exceptional conditions. Most are found as impressions in very fine-grained rock. Jellyfish, which are planktonic, are typically umbrella-shaped, and are fringed with tentacles hanging down from the edges of their bodies. The tentacles have stinging cells that paralyze their victims.

The *hydrozoans* are also rare as fossils. This is so because most of them live in fresh water and do not secrete any hard parts. A major exception is the order *stromatoporoidea,* an extinct group of hydrozoans that had a geologic time range of Cambrian through Cretaceous. The stromatoporoids (Figure V-4) were exclusively colonial marine animals that consisted of dense, calcareous laminated layers. During the Silurian and Devonian, the stromatoporoids were especially abundant, and constructed large reefs and reeflike structures.

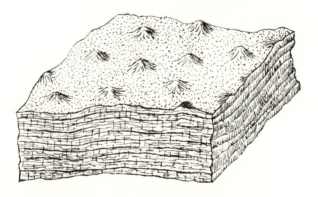

Figure V–4. Stromatoporoid.

The *anthozoans,* or corals (Figure V-5), are the most common of the coelenterates, so members of this phylum are the ones the collector is most likely to encounter. The corals are exclusively marine, benthonic, sessile, and filter feeding, and either solitary or colonial. The living animal consists of a body cavity with a mouth at the open end surrounded by tentacles that have stinging cells. The animals secrete exoskeletons of calcium carbonate that help ensure corals being found as fossils. Three orders of corals of special interest and importance to the fossil collector are the *tabulate, rugose,* and *scleractinia* corals.

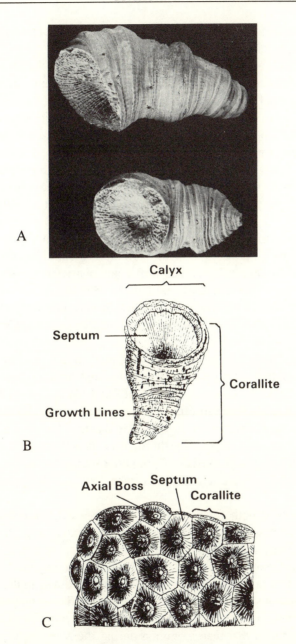

Figure V–5.   Coelenterata. (A) *Heliophyllum halli,* a solitary coral. (B) Morphology and principal parts of a solitary coral. (C) Typical colonial coral and its principal parts. (Photograph courtesy of Smithsonian Institution.)

The *tabulate corals* are an extinct group of fossil corals that had a geologic time range from the Ordovician through the Jurassic. Exclusively colonial, they were characterized by horizontal partitions called *tabulae*. They lacked the vertical radial partitioning characteristic of the rugose and scleractinian corals. The tabulates were important reef-builders in the Paleozoic—particularly in the Ordovician, Silurian, and Devonian Periods—and were presumably shallow-water dwellers.

The *rugose corals,* both solitary (sometimes called *horn corals*) and colonial, had a geologic time range of Ordovician through Permian. The rugose corals displayed a basic fourfold symmetry of septa (vertical partitions). The colonial forms were presumably warm, shallow-water dwellers, whereas the solitary forms may have lived in deeper waters. The former were important Paleozoic reef-builders.

The *scleractinian corals* also include both solitary and colonial forms. They had a geologic time range from the Triassic through the Present. They display a basic sixfold symmetry. Today, the colonial forms are restricted to warm, shallow waters, whereas some of the solitary forms have been found in both very deep and shallow water, and also in cold as well as warm water. The scleractinian corals were especially important as reef-builders during the Mesozoic and Cenozoic Eras.

Phylum Bryozoa: The *bryozoans* are strictly colonial animals that are structurally more complex than the coelenterates. The individual animals are very small, averaging less than 1 millimeter in length. Nonetheless, some of the colonies that they have constructed range up to 60 centimeters in diameter.

The bryozoans superficially resemble corals, in that they have a ring of tentacles surrounding a saclike body form. The bryozoans, however, possess U-shaped digestive tracts, each of which consists of a mouth, esophagus, stomach, intestine, and anus. Bryozoans also have separate sex organs, but they do not have hearts or vascular systems.

The bryozoans have both freshwater and marine-water representatives. The freshwater forms, however, have no hard parts and no fossil record. Therefore, you need be concerned only with the marine forms.

All bryozoans are benthonic, sessile, filter feeders, and they have a worldwide distribution. Most live in shallow water, and presumably did so in the past. Figure V-6 shows the different shapes one is likely to encounter. These include encrusting forms, which can be found covering shells, erect twiglike branching types, as well as lacy, fenestrate forms that were common during the Mississippian Period.

The geologic time range for the bryozoans was Ordovician through the Present.

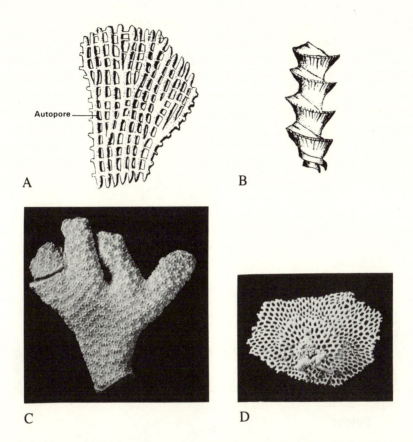

Figure V–6.   Bryozoa. (A) Fenestrate type. (B) *Archimedes,* a spiral type. (C) Twiglike branching type. (D) Encrusting type. (Photographs courtesy of Smithsonian Institution.)

Phylum Brachiopoda: *Brachiopods* are animals with bivalved shells that have a geologic time range of Cambrian through the Present. They were especially abundant during the Paleozoic Era. Consequently, they are one of the more common fossils you are likely to find in Paleozoic Age rocks. Because they were so diverse and abundant during the Paleozoic, they are very useful for age-dating and correlation of their containing rocks.

Brachiopods consist of a soft body interior enclosed by two calcareous or phosphatic valves or shells. The body consists of a digestive tract, a liver, genitals, a primitive vascular system without a heart, and a very simple nervous system.

They also have muscles to open and close the valves. Some brachiopods have a fleshy, muscular extension that comes out of the beak area; it is called a *pedicle* and is used to attach the animal to the sea floor.

The two valves of the brachiopod are unequal in size and shape, yet each valve is itself bilaterally symmetrical (Figure V-7). This means that if you draw a line from the beak to the back of the brachiopod, the left side of the shell (either shell!) is the mirror image of the right side.

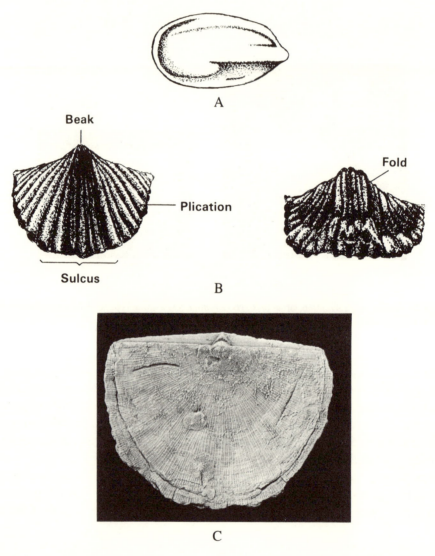

Figure V–7.   Brachiopods. (A) Inarticulate brachiopod. (B–D) Articulate brachiopods: (B) Morphology and principal parts. (C) *Rafinesquina ponderosa*. Also shows corals and bryozoans growing on its shell. *(continued on next page)*

D

(D) *Mucrospirifer arkonensis.* (Photographs courtesy of Smithsonian Institution.)

The brachiopods can be separated into two classes, the *inarticulate brachiopods* and the *articulate brachiopods*. This classification is based on the way the shells open and close.

The *inarticulate brachiopods*, the more primitive of the two classes, are characterized by having either a calcareous or chitinophosphatic shell. They lack hinged teeth, and open and close by muscles. The vast majority of inarticulate brachiopods are benthonic, sessile, epifaunal, filter feeders. Some, however, are infaunal filter feeders and use their pedicles to help them burrow into the sediment and anchor themselves there.

The *articulate brachiopods*, which are more common than the inarticulate brachiopods, possess calcareous valves and a tooth-and-socket arrangement where the two valves are hinged. All articulate brachiopods are benthonic, sessile, epifaunal, filter feeders. Most are found in a well-preserved state, with both valves together. These brachiopods underwent their greatest diversification during the Paleozoic Era. They make up only a minor element of today's marine invertebrate assemblage.

Phylum Mollusca: The phylum *Mollusca* contains a large group of terrestrial and aquatic organisms. In terms of number of species, the molluscs are the second largest living phylum. They typically display bilateral symmetry, a mantle or fleshy body covering, and calcareous external shells. Five classes are generally recognized. The *amphineura* or *chitons* are the least complex, followed by the *scaphopoda* or tusk shells. The other three classes are the rather familiar *gastropods* or snails, *pelecypods* or clams, and *cephalopods,* which include the squid and the octopus. We will discuss each of these classes in turn.

Class Amphineura: The *amphineura* have a geologic time range of Cambrian through the Present. The members of this class, commonly called *chitons,* consist of a shell composed of eight articulating plates that cover a fleshy foot (Figure V-8). Chitons are exclusively marine benthonic, mobile herbivores and, in spite of their long geologic history, are rare as fossils. This is so because they live in shallow water near shore, and, when an animal dies, its plates become separated and usually break up before being buried and subject to preservation.

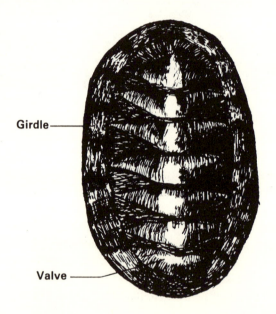

Girdle

Valve

Figure V–8.   Amphineura.

Class Scaphopoda: The *scaphopods* are commonly called tusk shells because they look like small tusks (Figure V–9). The animal consists of a slightly curved and tapering conical shell that is partly buried at an angle in the sea bottom. Their geologic time range is Silurian through the Present. They are easily identified, but may be mistaken for worm tubes.

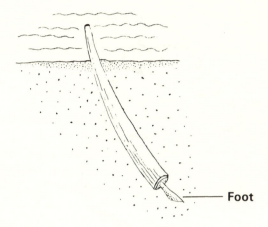

Figure V–9.   Scaphopoda.

Class Gastropoda: The *gastropods* constitute the largest class of molluscs, and include such diverse animals as snails, slugs, limpets, and abalones. Some gastropods are the only invertebrates, besides the insects and arachnids, to have adapted themselves to dry land.

Gastropods have a distinct head with a pair of eyes and tentacles, and move around on a broad, flat, muscular foot. Their shells exhibit a wide variety of sizes, shapes, and ornamentations; the common morphology for one is shown in Figure V-10. Most gastropod shells are coiled. To distinguish a gastropod shell from a coiled cephalopod shell, note that gastropod shells are not internally partitioned whereas cephalopod shells are.

A

Figure V–10.   Gastropoda. (A) *Worthenia tabulata*. (*continued on next page*)

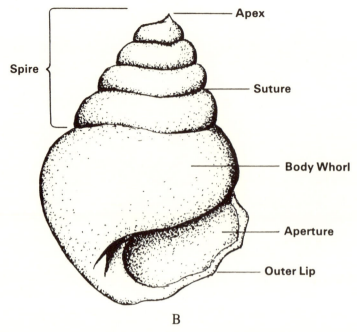

Apex

Spire

Suture

Body Whorl

Aperture

Outer Lip

B

(B) Morphology and principal parts of a gastropod. (Photograph courtesy of Smithsonian Institution.)

Most gastropods are benthonic, mobile herbivores that eat their food by scraping the surface of rocks or shells for algae. Some gastropods are benthonic, mobile carnivores that leave evidence of their activity in the form of small, neat, circular holes in the shells of other animals. Carnivorous gastropods drill a hole in the shell of their victim, inject a toxin that causes the muscles that hold the shells together to relax, and when the shells open up, then they eat. The geologic time range of the gastropods is Cambrian through the Present.

Class Pelecypoda: The *pelecypods* differ from other molluscs in that they have two calcareous shells that enclose the soft parts of the animal (Figure V-11). Pelecypods are aquatic and occur in fresh or marine water. Some pelecypods swim by rapidly opening and closing their shells. Most, however, are benthonic, sessile, epifaunal filter feeders, and are attached to the sea floor by thin organic strings called *byssal threads*. Some pelecypods move about on the sea floor through the use of a spade-shaped foot that they protrude from the shell and use to pull themselves forward. Others use their foot to burrow into the sediment, where they can live safe from predators. Some pelecypods even bore into solid rock or pilings and live their whole lives there.

A

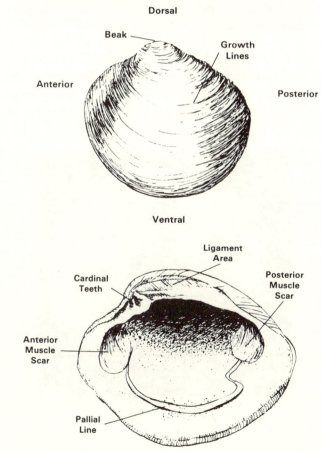

B

C

Figure V–11. Pelecypoda. (A) *Orthonota undulata*. (B, C) Morphology and principal parts of a typical pelecypod. (Photograph courtesy of Smithsonian Institution.)

The two shells (valves) of pelecypods, except for those of a few oysters, are of equal size and shape. This distinguishes them from the brachiopods, which have unequally sized shells. The shells are hinged by a tooth-and-socket arrangement and are held together by a ligament. The ligament spreads the valves apart, while a pair of muscles, when they contract, closes the shell. When a pelecypod dies, the closing muscles relax, thus springing the shells open. When the ligament rots, the shells come apart and may be moved around by currents. Consequently, unlike those of brachiopods, pelecypod shells are not usually found together.

Most pelecypods live in shallow marine waters, although many are found in deep water. They occur the world over. They are most abundant in Mesozoic and Cenozoic deposits. Their geologic time range is Cambrian through the Present.

Class Cephalopoda: The *cephalopods,* exclusively nektonic, carnivorous, marine animals, are the most highly developed of the molluscs. They have a distinct head with a pair of eyes and tentacles, some of which have sucker discs on them. They can move and maneuver by ejecting water through an opening at the bottom of their bodies. Cephalopods include forms both with and without shells, but only the shelled forms are at all common in the fossil record.

The shelled forms of cephalopods include both straight and coiled forms. The principal morphology is shown in Figure V-12. One of the interesting features of cephalopods is the way their shells are divided into compartments or chambers. The

A

Figure V–12. Cephalopoda. (A) *Imitoceral rotatorius.* (*continued on next page*)

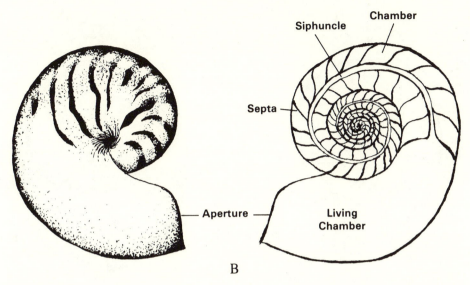

(B) Morphology and principal parts of a pearly *Nautilus*. (Photograph courtesy of Smithsonian Institution.)

junction between a partition and the outer shell is called the *suture*. The suture patterns of cephalopods range from simple to complex (Figure V-13). These patterns are not visible on the outside of the shells; they can be seen only when the outer layer of shell has been removed. The type of suture pattern is one way of classifying the cephalopods. Today only one shelled type, the pearly *Nautilus*, with a simple suture pattern, survives.

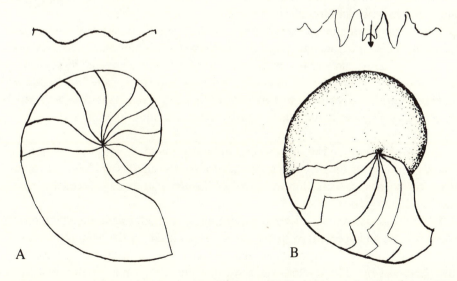

Figure V–13.   Cephalopod sutures. (A) Nautiloid type. (B) Goniatite type. (*continued on next page*)

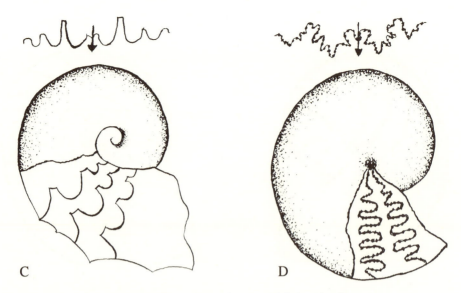

(C) Ceratite type. (D) Ammonite type. Arrow points toward living chamber.

The cephalopods are excellent fossils for age-dating and correlating of rocks, particularly in the Paleozoic and Mesozoic. Their geologic time range is Cambrian through the Present.

Phylum Annelida: The *annelid worms* include such familiar worms as the earthworm. The typical worm has a long, cylindrical, segmented body. Found in terrestrial, freshwater, and marine environments, worms have a geologic time range of Precambrian through the Present.

Worms usually are not preserved because they lack hard parts. When found in the fossil record, most are only impressions. Some forms, however, possessed jaws composed of a resistant organic material, and these can be recovered by dissolving the surrounding rock with various acids. These jaw structures, called *scolecodonts,* are microscopic in size. Some worms also produce calcareous tubes, some of which are found attached to brachiopods, pelecypods, and other hard objects.

Phylum Arthropoda: The *arthropods,* the most abundant and diverse group of animals known, include highly specialized animals such as spiders, scorpions, ticks, centipedes, insects, lobsters, crabs, shrimp, ostracodes, barnacles, trilobites, and eurypterids.

The arthropods are characterized by chitinous, jointed, bilaterally symmetrical exoskeletons, and specialized jointed appendages used for the capture of prey as well as for locomotion, respiration, and reproduction. As arthropods grow, they shed their previous exoskeletons and grow new ones. They may do this up to eight or more times before reaching adulthood. This may explain why we find so many parts of arthropods in the fossil record but few whole specimens.

Arthropods have successfully adapted to all environments, from the air to the land, and from freshwater to marine environments. From this standpoint, they are the most successful of all invertebrate animals. Their geologic time range is from Early Cambrian to the Present. In spite of their tremendous abundance and diversity, however, only a few forms are important as fossils. These are the *trilobites, eurypterids, crustaceans, ostracodes, barnacles*, and *insects*. We will take a closer look at each of these types of arthropods.

Class Trilobita: The *trilobites* are extinct, exclusively marine, benthonic, mobile detritus-deposit feeders (Figure V-14). They are one of the earliest (that is, most ancient) arthropods and ranged from the Early Cambrian to the end of the Permian.

A

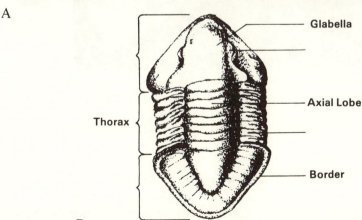

B

Figure V–14.   Trilobites. (A) *Phacops rana*. (B) Morphology and principal parts of a trilobite. (Photograph courtesy of Smithsonian Institution.)

Trilobites are bilaterally symmetrical and consist of a *cephalon* (head), *thorax* (body), and *pygidium* (tail). They attained their maximum abundance and diversity in the Cambrian and Ordovician, and then began to decline steadily in numbers until their extinction at the end of the Permian.

Most trilobites are found as parts, rather than as whole organisms. A few are found enrolled (that is, rolled up like a clenched fist). Trilobites are among the most popular and prized of fossils for amateur and professional collectors alike.

Class Meristomata: The *meristomes* are a large group of arthropods characterized by four pairs of appendages and specialized gills. This class includes the horseshoe crabs, scorpions, spiders, and eurypterids, the last an extinct order.

The *eurypterids,* which are extinct marine and freshwater arthropods, are sometimes referred to as *sea scorpions* (Figure V-15). They are characterized by large pincerlike claws. The eurypterids ranged from the Ordovician through the Permian and were most abundant during the Silurian and Devonian. Most eurypterids were relatively small, but some species attained lengths of more than two meters.

Figure V–15.   Eurypterid. *Eurypterus locustris.* (Photograph courtesy of Smithsonian Institution.)

Subphylum Crustacea: The *crustaceans* are characterized by a hard, calcite-impregnated exoskeleton consisting of a fused cephalothorax and abdomen. They also possess highly specialized appendages. Most crustaceans are aquatic. Several classes of crustaceans are recognized, including shrimps, crabs, lobsters, and copepods. They range from the Cambrian through the Present. Most have left a meager fossil record. For our purposes, the important ones are the *ostracodes* and the *barnacles*.

Class Ostracoda: The *ostracodes* are small, bivalved, aquatic crustaceans that are useful for age-dating and correlating rocks (Figure V-16). Externally they look like small clams, but they possess the typical segmented arthropod body and jointed appendages, which the shell covers and protects. The geologic time range of the ostracodes is Ordovician through the Present.

Figure V–16.   Ostracodes. (Photograph courtesy of Smithsonian Institution.)

Ostracodes, like foraminifers, are usually prepared as microfossils and studied under the microscope.

Class Cirripedia: The *cirripedia* include the exclusively marine, sessile *barnacles*. Externally, barnacles do not look like other arthropods, although they do possess the typical arthropod body and appendages. The body and the appendages, however, are housed in a calcareous exoskeleton consisting of several plates. Barnacles live attached to other objects and are commonly attached to other fossils. Their geologic time range is Ordovician through the Present (Figure V-17).

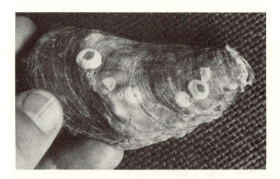

Figure V–17.   Barnacles. Present-day barnacles attached to a Cretaceous oyster shell. (Photograph courtesy of G. K. McCauley.)

Subphylum Insecta: The *insects* constitute the most diverse and abundant of all living invertebrates. Most are terrestrial and air-breathing, have six legs, flexible exoskeletons, and bodies divided into a head, a thorax, and an abdomen. Most also bear paired wings.

As you might expect from where insects live, they are very rare in the fossil record. Most insect fossils are found in amber or as impressions in very fine-grained rocks. They also occur replaced by silica; some of the finest examples of such replacement are the silicified insects from the Calico Mountains of southern California (Figure 4–7B).

The geologic time range of insects is Silurian through the Present. Some of the Pennsylvanian insects, which were extremely abundant, grew to be quite large. Impressions in coal layers of dragonflies with wing spans of two feet and cockroaches three and four inches long are not uncommon.

Phylum Echinodermata: *Echinoderms* are so named because of their typical "spiny skins," which are characteristic of animals of this phylum. Echinoderms also display fivefold (pentameral) symmetry and possess skeletons made up of numerous calcareous plates covered by a leathery outer skin. Echinoderms also possess a water vascular system, which is a well-developed hydrostatic pressure system that operates, for example, the tube feet of starfish.

Echinoderms are exclusively marine animals; their living habits range from those characteristic of epifaunal sessile filter feeders, to those of mobile carnivores, to those of infaunal mobile detritus-deposit feeders.

While there has been a great increase in knowledge about echinoderms in recent years and many different ways of classifying them have been put forth, their division into two subphyla based on mobile or sessile living habits remains the most practical for the collector. We shall, therefore, follow that classification. The important echinoderms are the *cystoids, blastoids, crinoids, holothurians, stelleroidians,* and *echinoids*.

Subphylum Pelmatozoa: These are echinoderms that are more or less permanently attached to the sea floor by either a stem or stalk. The pelmatozoans range from the Cambrian through the Present but were especially abundant during the Paleozoic Era. They include several classes. Only three are common and likely to be encountered by the average collector: the *cystoids,* the *blastoids,* and the *crinoids.*

Class Cystoidea: The *cystoids* are benthonic, sessile filter feeders that were relatively common during the Early Paleozoic (Figure V-18). They consist of a globular or saclike calyx (main body) composed of a number of irregularly arranged calcareous plates. The plates are typically perforated by pores or slits that probably served as sites for respiration. Cystoids had many small arms at the top of the calyx for food gathering and were attached to a substrate by a short stem.

The cystoids ranged from the Early Cambrian through the Devonian and were especially abundant in the Ordovician and Silurian.

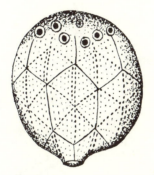

Figure V–18.   Cystoid.

Class Blastoidea: The *blastoids* are benthonic, sessile filter feeders (Figure V-19). Their calyxes consisted of 13 calcareous plates arranged in three rows of five, five, and three plates each. The blastoids had very short arms, which are only rarely preserved. They also had a short stem for attachment to the sea floor; this is also very rarely preserved.

Figure V–19.   Blastoid. *Pentremites pyriformis.* (Photograph courtesy of Smithsonian Institution.)

The blastoids had a geologic time range of Ordovician through Permian and were especially abundant during the Mississippian.

Class Crinoidea: The *crinoids*, the most important of the Paleozoic echinoderms, have gradually declined in both diversity and abundance right up to the present. Crinoids are usually benthonic, sessile filter feeders (Figure V-20). Some present-day forms, however, are not attached to the sea floor; instead they either are freely mobile or anchor themselves to loose objects by their short stems.

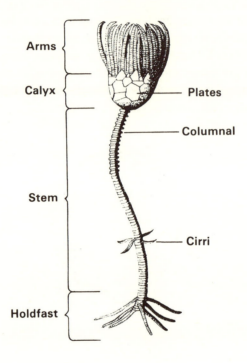

Figure V–20.   Crinoid. Shows morphology and principal parts.

The crinoid skeleton is composed of three parts: the stem, the calyx, and the arms. The stem is typically long in Paleozoic crinoids and short or nonexistent in today's species. It is rare to find the stem attached to the calyx in fossil specimens, because it was disarticulated so easily. However, individual pieces of the stems, called *columnals,* are common fossils in many rocks. The calyx, which encloses the animal's soft parts, consists of a series of calcareous plates arranged symmetrically in rows. The arms, used for food gathering, are generally not preserved. Individual crinoids commonly live in close proximity to other crinoids and are especially abundant in reefs. Consequently, the remains of several individuals are often found concentrated in small areas.

Their geologic time range is Ordovician through the Present, and, like blastoids, they were especially abundant in the Mississippian Period.

Subphylym Eleutherozoa: The *eleutherozoans* are benthonic, mobile echinoderms that are divided into several classes. The three most common are the *holothurians, stelleroids,* and *echinoids*. Only the echinoids are common as fossils.

Class Holothuroidea: The *holothuroids*, frequently called *sea cucumbers,* are benthonic, mobile detritus-deposit feeders that move on the sediment, ingesting it and extracting nutrients from it before excreting the residue.

The holothurians have an elongate, leathery, cucumber-shaped body with calcareous spicules scattered throughout the skin. The geologic time range of the holothurians is Ordovician through the Present.

Class Stelleroidea: The *stelleroids*, as their name implies, are star-shaped; they are also benthonic, mobile carnivores. They consist of a central disc from which five arms radiate outward. The arms have tube feet for attaching to objects and for holding onto bivalves while opening their shells to get to their soft parts. Both the central disc and the arms are covered by a leathery skin.

The stelleroids can be divided further into the *asteroids,* starfish (Figure V-21), and the *ophiuroids,* brittle stars. Neither starfish nor brittle stars are common in the fossil record; those found are generally preserved as impressions or molds.

Figure V–21.   Starfish. *Devonaster eucharis* (Reprinted by permission from R. V. Dietrich and B. J. Skinner, *Rocks and Rock Minerals*. © 1979 John Wiley & Sons, Inc., New York.)

The geologic time range of the stelleroids is Ordovician through the Present.

Class Echinoidea: The *echinoids,* the largest group of echinoderms, include the sea urchin, the heart urchin, and the sand dollar (Figure V-22).

A

B

Figure V–22. Echinoids. (A) *Stereocidaris hudspethensis,* a regular echinoid. (B) *Encope tamiamiensis,* an irregular echinoid. (Photographs courtesy of Smithsonian Institution.)

The echinoid body is composed of fused or articulating polygonal plates arranged in a series of radial rows. The exterior of the test is commonly covered with movable spines. The majority of echinoids are benthonic, mobile carnivores, although some have adapted to an infaunal existence.

The mouth is typically located on the bottom of the test and has a chewing or grinding apparatus, called *Aristotle's lantern,* for breaking up the shells of the animals it eats. It consists of five calcareous plates that are operated by a set of muscles.

There are two subclasses of echinoids based on the position of the anus. If the anus is located inside a ring of plates on the top of the animal, it is a *regular echinoid;* the sea urchin is a good example. If the anus is located outside the ring and is between that ring (on top of the shell) and the mouth (on the bottom), it is called an *irregular echinoid;* the heart urchin and the sand dollar are good examples.

The geologic time range of the echinoids is Ordovician through the Present. It wasn't until the late Mesozoic, however, that the echinoids became abundant and diverse. Many of these species are useful for age-dating and correlating Cretaceous and Tertiary rocks.

Phylum Protochordata: The protochordates possess a *notochord* (a long flexible rod, similar to a backbone) during at least part of their life cycle. The protochordates are intermediate between invertebrates and true chordates. The protochordate phylum includes acorn worms, sea squirts, and graptolites, an extinct group.

Class Graptolithina: *Graptolites* are a group of colonial animals that possessed a chitinous exoskeleton that housed the animal (Figure V-23). They are most commonly found as carbonaceous impressions in black shales, although they have also been found in other rocks.

Figure V–23.   Graptolite. *Tetragraptus*. (Photograph courtesy of Smithsonian Institution.)

Graptolites were apparently planktonic organisms either floating freely or attached to something that was floating. They had a worldwide distribution and the different species had short geologic ranges. Thus, they are excellent fossils for age-dating and correlation of rocks.

Their geologic time range was Cambrian through Mississippian. They were especially diverse and abundant during the Ordovician and Silurian.

Phylym Chordata: The *chordates,* the most complex of all animals, possess a notochord or backbone, a hollow dorsal nerve chord, and gill slits during at least part of their life cycle.

The chordates comprise many different classes and orders. We shall briefly discuss the more important ones: *fish, amphibians, reptiles, birds,* and *mammals.*

Superclass Pisces: Members of this superclass are the fishes, the simplest and most numerous of all vertebrates. Fishes are cold-blooded, aquatic, nektonic predators. There are four major classes of fish: the *agnathids,* the *placoderms,* the *chondrichthyes,* and the *osteichthyes.*

Agnathids are the most primitive fish. The earliest forms, known as *ostracoderms,* are found in Upper Cambrian rocks. They were jawless, lacked paired appendages, and were heavily armored (Figure V-24). These early ostracoderms were detritus-deposit feeders and swam on the bottom sucking food into a hole on the bottom of their head shield. The ostracoderms were abundant during the Ordovician and Silurian, declined during the Devonian, and became extinct by the end of the Devonian. Some members of the agnathids, however, evolved a parasitic life-style, and today are represented by the lamprey eel and hagfish, which do not have jaws, but lock onto the side of a fish and suck its blood.

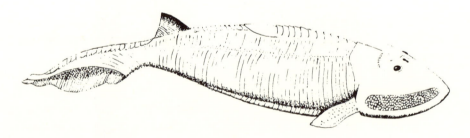

Figure V–24.   *Cephalaspis,* a Silurian ostracoderm.

The *placoderms* are an extinct group of jawed armored fish (Figure V-25). All placoderms had jaws, teeth, and paired appendages, and were covered with thick, bony armor. They first appeared in the Silurian, were the dominant fish in the Devonian, and became extinct near the end of the Permian. Certain species of placoderms grew to ten meters (~ 30 feet) in length during the Devonian. The placoderms are important because jaws, as we now know them, first appeared in this group; this effectively opened up a whole new niche, that of the carnivore.

Figure V–25.   *Dinichthys,* a Devonian placoderm.

The *chondrichthyes* are the cartilagenous fishes that include, for example, the sharks, rays, and skates. The chondrichthyes first appeared in the Devonian. The only parts of these fishes generally preserved as fossils are their teeth, which are relatively common as fossils in both Mesozoic and Cenozoic deposits.

The *osteichthyes* include all the bony fishes, which are the most highly developed and abundant fishes today. The general evolutionary trend of the bony fishes has been toward a more streamlined appearance, lighter skeleton, and more efficient jaws. Within this class are the *lungfishes,* which have developed the ability to breathe air when the water they are living in becomes foul or dries up. Related to the lungfishes and of great evolutionary importance are the *crossopterygians,* a group of fleshy finned fishes with lungs that evolved into amphibians in the Devonian.

Conodonts (Figure V-26) are also included in the superclass Pisces. Conodonts are microscopic, amber-colored, calcium phosphate toothlike structures. It is currently believed that conodonts made up some type of sieve structure for a now extinct group of fishes. Actually, they are an enigmatic group of fossils that have puzzled paleontologists for many years. They have a geologic time range of Ordovician through Triassic and, even though it is not known exactly what they are, they are very useful for age-dating and correlating rocks.

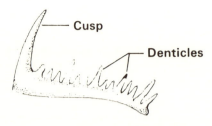

Figure V–26. Conodont. Shows morphology and principal parts.

Superclass Tetrapoda: The *tetrapods*, the most advanced chordates, are characterized by lungs, a three- or four-chambered heart, and paired appendages. Amphibians, reptiles, birds, and mammals are included in this superclass. We shall briefly discuss each class.

Class Amphibia: The *amphibians*, the earliest and most primitive of the tetrapods, are represented today by *frogs, toads, salamanders,* and *newts*.

The amphibians evolved from the crossopterygian fish in the Devonian and had many problems to overcome in making the transition from water to land. Among these problems were drying out, which they solved by developing a tough, watertight skin; the effect of gravity, which was solved by development of a strong backbone; the modification of fins to limbs; and reproduction, which was solved by returning to the water for laying of their eggs.

The early amphibians were called *labyrinthodonts* in reference to the complex infolding of their teeth (Figure V-27). There is an excellent fossil record of the labyrinthodonts. They first appeared in the Devonian, were abundant in the Late Pennsylvanian and Early Permian, and became extinct during the Triassic, when the more familiar amphibians became abundant. The labyrinthodonts were relatively short, squat animals that achieved a length of about two meters (about six feet); they undoubtedly ate heartily during the Pennsylvanian Period when insects in the coal-forming swamps were not only abundant but large.

Figure V–27. *Eryops,* a Permian labyrinthodont amphibian.

Class Reptilia: The *reptiles* are a large and varied group of cold-blooded, lung-breathing vertebrates, with an advanced, sturdy, bony skeleton. The major advance of the reptiles over the amphibians was the development of the amniote egg, which freed them from having to return to the water to lay their eggs. The amniote egg, a self-contained system that does not dry out, was a major evolutionary advance and allowed the reptiles to adapt to essentially all terrestrial environments. In fact, the Mesozoic Era is referred to as the "Age of Reptiles."

The reptiles have been divided into numerous subclasses, orders, and suborders by specialists. Only the more important groups are discussed here (Figure V-28).

The *cotylosaurs* are sometimes referred to as the *stem reptiles* because they are the most ancient and most primitive of the reptiles and apparently gave rise to all other reptile groups. They are characterized by a relatively small size, sprawling legs, and few bones in their skeleton. They had a geologic time range of Pennsylvanian through the Triassic.

The *pelycosaurs* or *sailback reptiles* are a group of primitive, mammal-like reptiles that lived from the Pennsylvanian through the Permian. They were anywhere from two to four meters ($\sim$ 6 to 12 feet) long and were characterized by a large fin or sail-like structure on their back. This sail is now believed to represent a type of temperature-regulating device. There were both herbivorous and carnivorous pelycosaurs.

The *therapsids* are the advanced, mammal-like reptiles that gave rise to the mammals during the Triassic. They developed many mammal-like features, as evidenced by their teeth, skulls, and skeletons. Instead of having their legs sprawled out on their sides like most other reptiles', the therapsids had their legs underneath their bodies, thus raising their bodies off the ground, just as the mammals do. The therapsids ranged from Middle Permian through the Triassic.

The *lizards* and *snakes* are a varied group. The lizards first appeared in the Triassic. The snakes are first known from Cretaceous rocks. Both groups are quite successful today.

The *turtles* and *tortoises* are another group of highly successful reptiles. They first appeared in the Triassic, and their major changes since then have been a reduction or elimination of teeth, and the development of the ability to retract their heads and legs into their shells.

After the transition to land and the development of the amniote egg, a whole group of reptiles returned to the sea. Then, the problems of the sea had to be resolved by evolutionary adaptations of the current reptilian characteristics. This illustrates the point that there is no retrogressive evolution as such; instead there are modifications of existing characteristics to resolve existing problems. This is exactly what happened when reptiles returned to the sea.

The *ichthyosaurs* were carnivorous, streamlined reptiles that superficially resembled porpoises or dolphins. They were three to five meters ($\sim$ 9 to 16 feet) long on the average, with some attaining a length of nearly ten meters ($\sim$ 30 feet). That ichthyosaurs were carnivores is evident from their large eyes and sharp teeth. The ichthyosaurs hatched their young internally from eggs carried by the female. The geologic time range of the ichthyosaurs was Triassic through the Cretaceous.

The *plesiosaurs* were long-necked, carnivorous, marine reptiles, attaining lengths up to 17 meters (~ 55 feet). They had squat bodies with paddlelike fins that enabled them to move forward as well as backward. The plesiosaurs also were fish eaters and had very long, sharp overlapping teeth. The geologic time range of the plesiosaurs was Triassic through the Cretaceous.

The *mososaurs* were carnivorous, marine lizards that attained a length of 10 meters (~ 30 feet). They had long, flexible bodies, powerful tails, and sharp teeth. Their main diet appears to have been cephalopods. They lived only during the Cretaceous.

The *placodonts* were very ugly marine reptiles that had a short, squat body, short legs, and a long tail. They swam around near shore and fed on beds of molluscs, using their short flat teeth. The placodonts lived only during the Triassic.

Other reptiles of the Mesozoic Era include the following.

The *thecodonts,* which only lived during the Triassic, were important because they gave rise to the phytosaurs, crocodilians, pterosaurs, dinosaurs, and perhaps also the birds. There were two groups of thecodonts. One was quadrapedal and heavy-boned. The other was bipedal, small, and thin-boned.

The *phytosaurs* were reptiles resembling crocodilians that lived only during the Triassic. They ranged in size from two to about eight meters (~ 6 to 25 feet) in length and had very sharp teeth, five clawed toes per foot, and powerful, long tails. Predaceous carnivores, the phytosaurs lived along the shores of streams, rivers, and lakes. They are distinguished from true crocodilians by the fact they had their nostrils on raised bumps in front of their eyes.

The *crocodilians,* which include crocodiles, alligators, and gavials, replaced the phytosaurs and took over their carnivorous-predator niche. They are distinguished from the phytosaurs by having their nostrils at the end of their snout. Their geologic time range is Triassic through the Present.

The *pterosaurs* have been referred to as flying reptiles. This is only partially true. One group (the *Rhamphorinchoids*) were small and could fly; the other group (the *Pterodactyloids*), some of which had wing spans of 16 meters (~ 50 feet) were gliders. The pterosaurs, insect and fish eaters, spent most of their time in the air; on the ground, they were extremely vulnerable to predators, as they had weak back legs. There is evidence that they may have been covered with fur, even though their body was definitely reptilian. Their geologic time range was Triassic through the Cretaceous.

The *dinosaurs* can be divided into two major groups on the basis of their hip structures. The *saurischians* had a lizardlike hip with a forward-directed pubis bone. There are two main groups of saurischians, the *theropods* and the *sauropods* (Figure V-28).

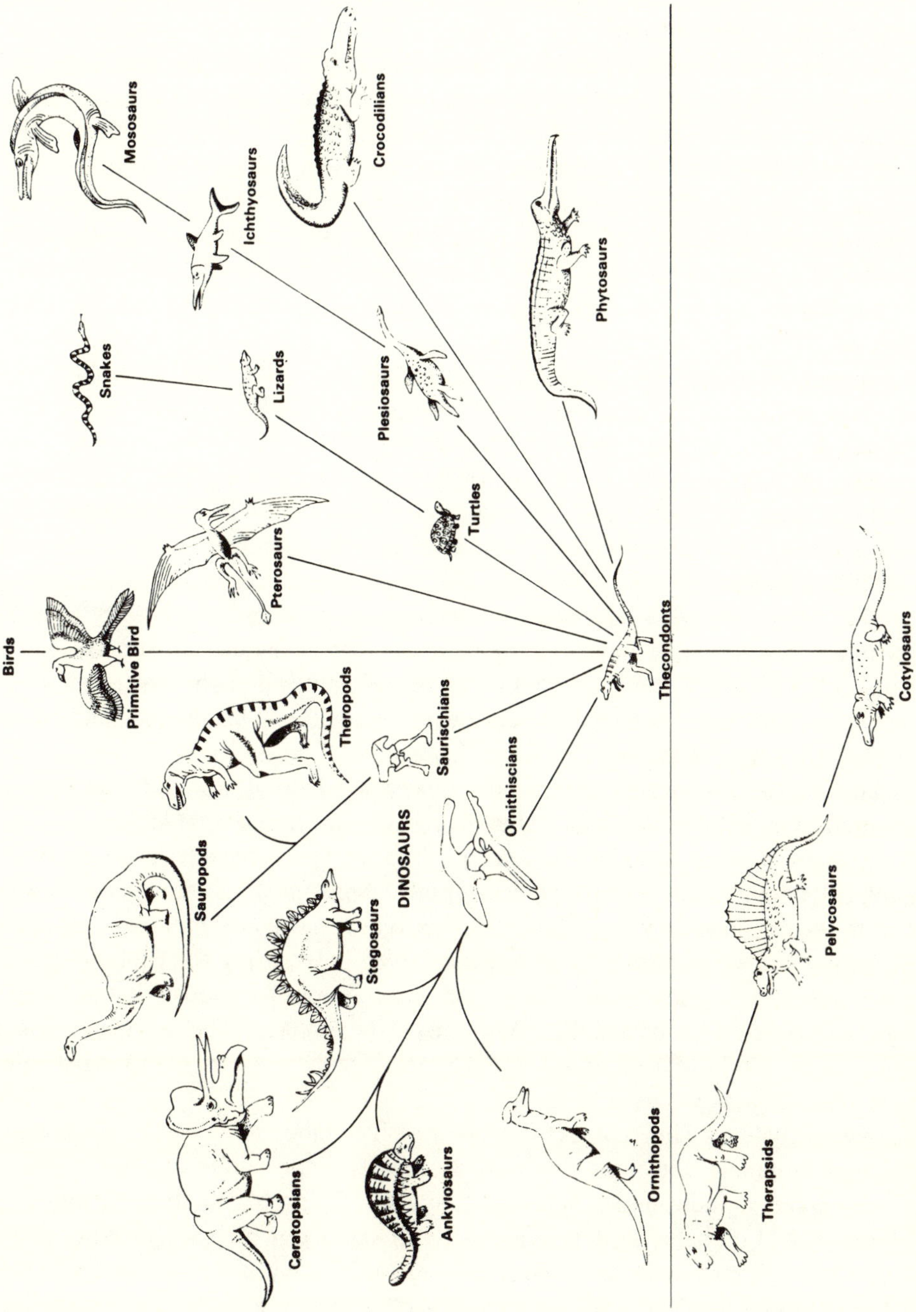

Figure V–28.   Reptile and bird evolutionary chart.

The *theropods* were bipedal carnivores. They ranged from the Triassic through the Cretaceous. The largest of the theropods were seven meters (~ 22 feet) tall up to 17 meters (~ 55 feet) long, and weighed several tons. They had sharp claws, and long, sharp, daggerlike teeth.

The *sauropods* were the largest of all dinosaurs, most reaching lengths of 27 meters (~ 90 feet) and weighing 30 to 40 tons. Recent finds, in fact, indicate a species of sauropod that was about 40 meters (~ 130 feet) long! They were quadrapeds with very long necks and tails and very small heads in relation to their bodies. In spite of their size, they were all herbivores. Recent research indicates the sauropods did not live in water, using their necks as periscopes, as indicated in some of the older books; instead, it appears that, when they went into the water, they probably swam. Their geologic time range was Triassic through the Cretaceous.

The *ornithiscians*, the other major group of dinosaurs, had a birdlike hip structure and are subdivided into four subgroups: the *ornithopods*, the *stegosaurs*, the *ankylosaurs*, and the *ceratopsians* (Figure V-28).

The *ornithopods* were the most primitive of the ornithiscians. They had a geologic time range of Triassic through the Cretaceous. The ornithopods were bipedal, semiaquatic herbivores. Some had large crests on their heads; a few had ducklike mouths.

The *stegosaurs* were herbivorous, quadrapedal animals, ranging up to eight meters (~ 25 feet) long and weighing up to ten tons. They had small heads and a series of bony plates on their backs. It was originally thought that these were defensive structures, but recently discovered evidence indicates that they were primarily a temperature-regulating mechanism.

The *ankylosaurs* were herbivorous, quadrapedal dinosaurs with flat bodies covered by a thick bony armor. They lived only during the Cretaceous.

The *ceratopsians*, also herbivorous quadrapeds, were characterized by a bony neck frill or shield that bore many sharp spines. Apparently they traveled in herds, eating vegetation as they went. The first discovered dinosaur eggs came from a nest of ceratopsian dinosaurs. The ceratopsians lived only during the Cretaceous.

Class Aves: Birds are seldom preserved as fossils because of the fragile nature of their skeletons and the environments they inhabit. The remains of some birds, however, have been found, and we can piece their history together from the sparse collections we have.

The oldest known bird comes from the Jurassic Solnhofen Limestone of Bavaria. Several complete skeletons, as well as feather impressions of this creature, *Archaeopteryx*, indicate that it was a true bird (Figure V-28). It could not fly, however, and probably used its wings only to knock down insects or as a net for catching small mammals. Its remains also indicate that it was reptilian in character, which supports the generally held hypothesis that there is a very close relationship between birds and reptiles.

The fossil record of the birds for the Cretaceous is more complete and indicates that there were true flying birds by that time. After the dinosaurs became extinct, at the end of the Cretaceous, it is apparent that the birds became even more diverse and, in fact, almost became the dominant large animal. During the Eocene, there were many large, flightless birds that effectively filled the large predatory-animal niche left vacant by the dinosaurs.

Since the Eocene, birds have continued to diversify and today live in essentially all environments and geographic areas.

Class Mammalia: Mammals are characterized by the fact that they are warm-blooded, have a protective covering of hair, breathe air, are born alive, and are fed with milk from the female's mammary glands.

The first mammals appear to have originated in the Triassic Period; they diversified rapidly during the Mesozoic Era, and proliferated in the Cenozoic Era, after the dinosaurs became extinct.

Mammalian classification recognizes several subclasses, orders, and suborders. Coverage of the mammals in a book of this nature, however, must be brief, so only the more important groups are described.

The *allotherians* were "Mesozoic mammals" that originated in the Triassic and became extinct in the Eocene. They were the ancestors of modern mammals. All of them were small, no larger than a modern house cat.

The *monotremes* are a group of bizarre mammals living today in Australia. The only two members of the monotremes are the spiny anteater and the duck-billed platypus. The monotremes are warm-blooded and have fur and mammary glands, but they lay eggs instead of bearing their young live. After the eggs hatch, however, the young are cared for in the same manner as other mammals. The monotremes, although found in Pleistocene deposits as fossils, are generally believed to have essentially descended unchanged from Mesozoic mammals.

The *therians,* which evolved in the Late Mesozoic, are represented by two groups, the *marsupials* and the *placentals*. The *marsupials,* the less successful of the two, constitute only about five percent of the Cenozoic mammals. They differ from the placentals chiefly in their method of reproduction. The marsupials have a short gestation period and are born at a very early state of development. The immature animals crawl to their mother's pouch, where they attach themselves to a nipple and complete their development there. The placental mammals, on the other hand, have a long gestation period and are born fully developed. The marsupials are most diverse and abundant in South America and Australia. They present good evidence of *parallel evolution;* that is, they provide examples of animals that perform the same functions and fill the same niches as their placental counterparts do in other places. Familiar marsupials include the kangaroo, the koala bear, and the opossum.

The *placentals* are the largest and most diverse of the mammal groups. During the Paleocene and Eocene, the placentals underwent explosive evolutionary radiation that led to the great diversity of mammals we see today.

The *carnivores* are fur-covered, flesh-eating animals, characterized by well-developed sharp teeth and clawed feet for tearing and slashing. The most ancient known carnivores appeared in the Paleocene and were about the size of sheep. The carnivores diversified during the Cenozoic Era. One of the more interesting carnivores was the saber-toothed tiger, which took the development of its canine teeth to an amazing extreme (Figure V-29).

Figure V–29.   Saber-toothed tiger, a Pleistocene carnivore.

The *edentata,* primitive herbivores characterized by poorly developed teeth, include the anteater, the tree sloth, and the armadillo. Two groups that are especially interesting are the giant ground sloths and the glyptodonts, both of which are now extinct (Figure V-30). Most of the remains of both of these mammals have been found in South America and the southern part of North America.

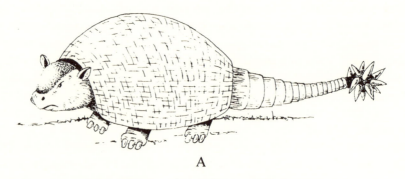

A

Figure V–30.   Edentata. (A) Glyptodont. (*Continued on next page*)

B

(B) Giant ground sloth.

The *proboscidians* are the elephants, mammoths, and mastodons, all of which are characterized by their large size, their simple teeth, and the elongation of their noses to trunks. The woolly mammoths are a group of extinct proboscidians found in Pleistocene deposits, including glacial ice. They had a long, thick, hairy coat and long, curved tusks (Figure V-31).

Figure V–31.   Proboscidian. Woolly mammoth.

The *artiodactyls* are even-toed, hoofed mammals, such as pigs, sheep, deer, camels, giraffes, goats, and hippopotami. The artiodactyls, obviously a large and varied group of mammals, have been around since the Eocene Epoch.

Figure V–32.   Brontothere.

The *perissodactyls* are odd-toed, hoofed animals, such as horses and rhinoceroses, as well as several extinct forms, such as the brontotheres (Figure V-32) and uintotheres (Figure V-33). The history of the horse is especially well known and documented (Figure 4–13).

Figure V–33.   Uintathere.

The *primates* include lemurs, monkeys, and apes, as well as humans. All are characterized by hands with opposable thumbs and stereoscopic vision. They are either tree dwellers or erect bipedal walkers.

PLANT DIVISIONS

The plant kingdom has left an abundant and diverse fossil record, not only in the form of plant remains, but also through spores and pollen. Plants are photosynthetic (that is, they must live where there is light), produce their own food, and, of course, are important in both the aquatic and terrestrial environments. Most amateur collectors do not normally collect plant remains. And even the plants that are collected tend to be poorly preserved, fragmental, and difficult to identify. Consequently, we discuss the major plant groups only briefly.

The *algae* have a long geologic history ranging from the Precambrian through the Present. Botanists divide the algae into many groups, such as the *diatoms* (Figure V-34), *dinoflagellates, coccolithophorids, green algae,* and *blue–green algae*. Paleontologists usually collect and prepare them by using specialized micropaleontologic techniques. Some algae secrete calcareous exoskeletons and have been important reef-formers.

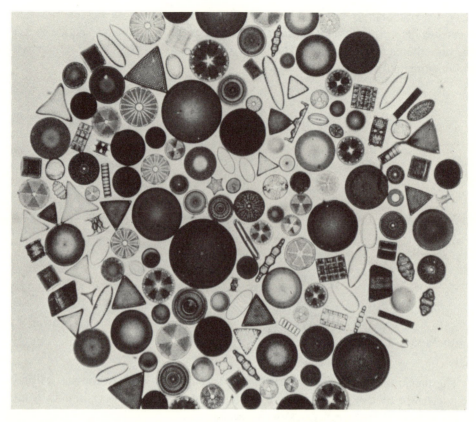

Figure V–34. Diatoms. (Photograph courtesy of M. H. Hohn.)

The *fungi* are similar to algae, but lack chlorophyll (the pigment that allows plants to photosynthesize). Thus, most forms depend on living or dead organic matter for their food. Fungi are rare in the fossil record.

The *bryophytes* are simple plants that lack roots and vascular tissue. They include the mosses and liverworts. They are extremely rare in the fossil record.

The *tracheophytes* include all plants that possess vascular tissue. The three major groups of tracheophytes are the *seedless vascular plants,* the *gymnosperms,* and the *angiosperms*. The only unifying characteristic of each group is the method of reproduction; consequently, this characteristic also constitutes the major means of separating the groups from one another.

The *seedless vascular plants* are the simplest (Figure V-35). They comprise the ferns, which reproduce like many green algae (that is, they alternate a sexual and asexual generation). They are restricted to moist areas such as forests and swamps because of the need for moisture during the reproductive cycle. The time of the greatest diversity and abundance of the seedless vascular plants was during the Pennsylvanian Period, when much of the coal we use today was forming in large swampy areas.

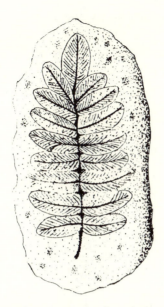

Figure V–35.   Seedless vascular plant. Impression in an ironstone concretion.

The *gymnosperms* are mostly evergreens, which bear naked seeds. Most have a large, dominant trunk and many smaller branches. The major groups are the *cycads,* which look like palm trees, the *ginkgos* or *maidenhair trees* (Figure V-36), and the *conifers,* which are the largest group of gymnosperms. The conifers include pine, spruce, fir, juniper, and redwood.

Figure V–36.   Gymnosperm. Ginkgo leaves.

The *angiosperms*, or flowering plants (Figure V-37), include all of the familiar flowers, as well as grasses, fruits, vegetables, and hardwoods. This group, which has adapted itself to essentially all environments, includes over 250,000 living species as well as a large number of fossil forms.

Figure V–37.   Angiosperm.

APPENDIX VI

Chemical Symbols and Formulas

Each chemical element has been given a one- or two-letter symbol that has been accepted and is used in, for example, chemical formulas throughout the world. Symbols for the elements that are in the formulas of relatively common minerals are given in Table VI–1.

The following description of chemical formulas, by John Sampson White of the Smithsonian Institution, is reprinted here by kind permission of the author and Michael Fleisher, who holds the copyright. The paragraph is from: Michael Fleischer, *Glossary of Mineral Species, 1980*. Mineralogical Record, Tuscon, Ariz. 1980.

In order to learn how to read a formula, one must first realize that there are two basic types of components—cations and anions (including anionic groups). The former are always positive (+) charged, the latter negative (–). Cations are also always written first in a formula; that is, they appear on the left, anions on the right. There may be only one or several of each. In **cubanite**, $CuFe_2S_3$, for example, copper (Cu) and iron (Fe) are cations and sulfur (S) is an anion. The mineral is a copper iron sulfide. Anionic groups are composed of a positively charged element combined with oxygen such that they behave and may be thought of as single negatively charged entities. Examples are sulfate (SO_4^{-2}), silicate (SiO_4^{-4}), phosphate (PO_4^{-3}), and carbonate (CO_3^{-2}). Anionic groups are often enclosed in parentheses but they need not be. Berlinite, $AlPO_4$, and beusite, $(Mn,Fe,Ca,Mg)_3(PO_4)_2$, are both phosphates. . . . a succession of elements within parentheses, separated by commas, signifies that mutual substitution of all of them is known to occur but the first (on the left) is the most abundant and, perhaps, the only essential one within the parentheses. On the opposite end of the formula we often see two or more anionic groups plus both hydroxyl (OH)⁻ and water (H_2O). All are essential if they are presented each in its own set of parentheses. For example, **chalcophyllite**, $Cu_{18}Al_2(AsO_4)_3(SO_4)_3(OH)_{27} \cdot 33H_2O$, may be phrased as copper aluminum arsenate sulfate hydroxide hydrate. A few additional examples will serve to illustrate just how logical this really is. **Forsterite**, Mg_2SiO_4, is magnesium silicate. **Rhodonite,**

$(Mn,Fe,Ca,Mg)SiO_3$, is manganese silicate. Iron, calcium, and magnesium are known to substitute for manganese in rhodonite, but manganese must be dominant. If iron or calcium or magnesium were dominant, and listed ahead of manganese, the mineral would not be rhodonite. Some minerals, such as many of the zeolite group experience mutual aluminum and silicon substitution. Thus, for convenience, they are usually referred to as aluminosilicates. **Stilbite**, $NaCa_2(Al_5Si_{13})O_{36} \cdot 14H_2O$, is expressed as sodium calcium aluminosilicate hydrate.

Table VI–1 Chemical Elements and Anionic Groups That Are in Formulas of Relatively Common Minerals.*

Chemical Elements				Anionic Groups	
Ag	Silver	Mo	Molybdenum	AsO_4	Arsenate
Al	Aluminum	N	Nitrogen	BO_3	Borate
As	Arsenic†	Na	Sodium	CO_3	Carbonate
Au	Gold	Nb	Columbium	CrO_4	Chromate
B	Boron	Ni	Nickel	MoO_4	Molybdate
Ba	Barium	O	Oxygen†	NO_3	Nitrate
Be	Beryllium	P	Phosphorus	OH	Hydroxyl
C	Carbon	Pb	Lead	PO_4	Phosphate
Ca	Calcium	Pt	Platinum	SO_4	Sulfate
Ce	Cerium	S	Sulfur†	SiO_4	
Cl	Chlorine†	Si	Silica	Si_2O_7	
Co	Cobalt	Sn	Tin	SiO_3	Silicates
Cr	Chromium	Sr	Strontium	Si_4O_{11}	
Cu	Copper	Ta	Tantalum	Si_2O_5	
F	Fluorine†	Th	Thorium	UO_2	Uraninate
Fe	Iron	Ti	Titanium	VO_4	Vanadate
H	Hydrogen	U	Uranium	WO_4	Tungstate
Hg	Mercury	V	Vanadium		
K	Potassium	W	Tungsten		
Li	Lithium	Zn	Zinc		
Mg	Magnesium	Zr	Zirconium		
Mn	Manganese				

*To facilitate reference, the elements are listed alphabetically according to their symbols and the anionic groups are listed alphabetically by the first letter of each group.

†These elements typically act as anions and the resulting compounds are called, for example, oxides, fluorides, and sulfides.

INDEX

Asterisks(*) indicate illustrations. Mineral and rock entries in the determinative tables (p. 120–156) are not included.